THE
BODY
ELECTRIC

THE BODY ELECTRIC

Electromagnetism and the Foundation of Life

ROBERT O. BECKER, M.D.,
and GARY SELDEN

Illustrated by David Bichell

WILLIAM MORROW
An Imprint of HarperCollins*Publishers*

HB 03.15.2024 1035

Library of Congress Cataloging-in-Publication Data

Becker, Robert O.
  The body electric.

  Reprint. Originally published: New York: Morrow, ©1985.
  Includes index.
  1. Electromagnetism—Physiological effect.
2. Electromagnetism in medicine.   I. Selden, Gary.
II. Title.
QP82.2.E43B4   1987   591.19′127   86-25168
ISBN 0-688-06971-1 (pbk.)

Printed in the United States of America

BOOK DESIGN BY PATTY LOWY

24 25 26 27 28 LBC 77 76 75 74 73

To my wife, Lillian

—R.O.B.

To Harwood Rhodes, from a grateful student

—G.S.

Acknowledgments

We wish to thank our wives, Lillian Becker and Maureen Sugden, without whose love, help, and patience this book would never have been completed. We also wish to acknowledge the contributions of editor Maria Guarnaschelli and copy editor Bruce Giffords, as well as Julie Weiner, the editor who first saw promise in this project, and Susan Schiefelbein, who began a draft of the work several years ago. In addition, we are grateful to friends, colleagues, researchers, and sources too numerous to list. To those not mentioned in the text we hereby offer our heartfelt thanks.

Co-author's Note

Bob Becker spent almost thirty years working on the substance of this book. I spent not quite two in helping to organize the material and fit the words together. Therefore I have chosen to tell the entire story from his point of view. Unless otherwise noted, "I" refers to him and "we" to him and his collaborators in research.

— GARY SELDEN

Contents

THE
BODY
ELECTRIC

Introduction:
The Promise of the Art

I remember how it was before penicillin. I was a medical student at the end of World War II, before the drug became widely available for civilian use, and I watched the wards at New York's Bellevue Hospital fill to overflowing each winter. A veritable Byzantine city unto itself, Bellevue sprawled over four city blocks, its smelly, antiquated buildings jammed together at odd angles and interconnected by a rabbit warren of underground tunnels. In wartime New York, swollen with workers, sailors, soldiers, drunks, refugees, and their diseases from all over the world, it was perhaps *the* place to get an all-inclusive medical education. Bellevue's charter decreed that, no matter how full it was, every patient who needed hospitalization had to be admitted. As a result, beds were packed together side by side, first in the aisles, then out into the corridor. A ward was closed only when it was physically impossible to get another bed out of the elevator.

Most of these patients had lobar (pneumococcal) pneumonia. It didn't take long to develop; the bacteria multiplied unchecked, spilling over from the lungs into the bloodstream, and within three to five days of the first symptom the crisis came. The fever rose to 104 or 105 degrees Fahrenheit and delirium set in. At that point we had two signs to go by: If the skin remained hot and dry, the victim would die; sweating meant the patient would pull through. Although sulfa drugs often were effective against the milder pneumonias, the outcome in severe lobar pneumonia still depended solely on the struggle between the infection and the patient's own resistance. Confident in my new medical knowledge, I was horrified to find that we were powerless to change the course of this infection in any way.

It's hard for anyone who hasn't lived through the transition to realize the change that penicillin wrought. A disease with a mortality rate near 50 percent, that killed almost a hundred thousand Americans each year, that struck rich as well as poor and young as well as old, and against which we'd had no defense, could suddenly be cured without fail in a few hours by a pinch of white powder. Most doctors who have graduated since 1950 have never even seen pneumococcal pneumonia in crisis.

Although penicillin's impact on medical practice was profound, its impact on the philosophy of medicine was even greater. When Alexander Fleming noticed in 1928 that an accidental infestation of the mold *Penicillium notatum* had killed his bacterial cultures, he made the crowning discovery of scientific medicine. Bacteriology and sanitation had already vanquished the great plagues. Now penicillin and subsequent antibiotics defeated the last of the invisibly tiny predators.

The drugs also completed a change in medicine that had been gathering strength since the nineteenth century. Before that time, medicine had been an art. The masterpiece—a cure—resulted from the patient's will combined with the physician's intuition and skill in using remedies culled from millennia of observant trial and error. In the last two centuries medicine more and more has come to be a science, or more accurately the application of one science, namely biochemistry. Medical techniques have come to be tested as much against current concepts in biochemistry as against their empirical results. Techniques that don't fit such chemical concepts—even if they seem to work—have been abandoned as pseudoscientific or downright fraudulent.

At the same time and as part of the same process, life itself came to be defined as a purely chemical phenomenon. Attempts to find a soul, a vital spark, a subtle something that set living matter apart from the nonliving, had failed. As our knowledge of the kaleidoscopic activity within cells grew, life came to be seen as an array of chemical reactions, fantastically complex but no different in kind from the simpler reactions performed in every high school lab. It seemed logical to assume that the ills of our chemical flesh could be cured best by the right chemical antidote, just as penicillin wiped out bacterial invaders without harming human cells. A few years later the decipherment of the DNA code seemed to give such stout evidence of life's chemical basis that the double helix became one of the most hypnotic symbols of our age. It seemed the final proof that we'd evolved through 4 billion years of chance molecular encounters, aided by no guiding principle but the changeless properties of the atoms themselves.

The philosophical result of chemical medicine's success has been belief in the Technological Fix. Drugs became the best or only valid treatments for all ailments. Prevention, nutrition, exercise, lifestyle, the patient's physical and mental uniqueness, environmental pollutants—all were glossed over. Even today, after so many years and millions of dollars spent for negligible results, it's still assumed that the cure for cancer will be a chemical that kills malignant cells without harming healthy ones. As surgeons became more adept at repairing bodily structures or replacing them with artificial parts, the technological faith came to include the idea that a transplanted kidney, a plastic heart valve, or a stainless-steel-and-Teflon hip joint was just as good as the original—or even better, because it wouldn't wear out as fast. The idea of a bionic human was the natural outgrowth of the rapture over penicillin. If a human is merely a chemical machine, then the ultimate human is a robot.

No one who's seen the decline of pneumonia and a thousand other infectious diseases, or has seen the eyes of a dying patient who's just been given another decade by a new heart valve, will deny the benefits of technology. But, as most advances do, this one has cost us something irreplaceable: medicine's humanity. There's no room in technological medicine for any presumed sanctity or uniqueness of life. There's no need for the patient's own self-healing force nor any strategy for enhancing it. Treating a life as a chemical automaton means that it makes no difference whether the doctor cares about—or even knows—the patient, or whether the patient likes or trusts the doctor.

Because of what medicine left behind, we now find ourselves in a real technological fix. The promise to humanity of a future of golden health and extended life has turned out to be empty. Degenerative diseases—heart attacks, arteriosclerosis, cancer, stroke, arthritis, hypertension, ulcers, and all the rest—have replaced infectious diseases as the major enemies of life and destroyers of its quality. Modern medicine's incredible cost has put it farther than ever out of reach of the poor and now threatens to sink the Western economies themselves. Our cures too often have turned out to be double-edged swords, later producing a secondary disease; then we search desperately for another cure. And the dehumanized treatment of symptoms rather than patients has alienated many of those who *can* afford to pay. The result has been a sort of medical schizophrenia in which many have forsaken establishment medicine in favor of a holistic, prescientific type that too often neglects technology's real advantages but at least stresses the doctor-patient relationship, preventive care, and nature's innate recuperative power.

The failure of technological medicine is due, paradoxically, to its success, which at first seemed so overwhelming that it swept away all aspects of medicine as an art. No longer a compassionate healer working at the bedside and using heart and hands as well as mind, the physician has become an impersonal white-gowned ministrant who works in an office or laboratory. Too many physicians no longer learn from their patients, only from their professors. The breakthroughs against infections convinced the profession of its own infallibility and quickly ossified its beliefs into dogma. Life processes that were inexplicable according to current biochemistry have been either ignored or misinterpreted. In effect, scientific medicine abandoned the central rule of science—revision in light of new data. As a result, the constant widening of horizons that has kept physics so vital hasn't occurred in medicine. The mechanistic assumptions behind today's medicine are left over from the turn of the century, when science was forcing dogmatic religion to see the evidence of evolution. (The reeruption of this same conflict today shows that the battle against frozen thinking is never finally won.) Advances in cybernetics, ecological and nutritional chemistry, and solid-state physics haven't been integrated into biology. Some fields, such as parapsychology, have been closed out of mainstream scientific inquiry altogether. Even the genetic technology that now commands such breathless admiration is based on principles unchallenged for decades and unconnected to a broader concept of life. Medical research, which has limited itself almost exclusively to drug therapy, might as well have been wearing blinders for the last thirty years.

It's no wonder, then, that medical biology is afflicted with a kind of tunnel vision. We know a great deal about certain processes, such as the genetic code, the function of the nervous system in vision, muscle movement, blood clotting, and respiration on both the somatic and the cellular levels. These complex but superficial processes, however, are only the tools life uses for its survival. Most biochemists and doctors aren't much closer to the "truth" about life than we were three decades ago. As Albert Szent-Györgyi, the discoverer of vitamin C, has written, "We know life only by its symptoms." We understand virtually nothing about such basic life functions as pain, sleep, and the control of cell differentiation, growth, and healing. We know little about the way every organism regulates its metabolic activity in cycles attuned to the fluctuations of earth, moon, and sun. We are ignorant about nearly every aspect of consciousness, which may be broadly defined as the self-interested integrity that lets each living thing marshal its responses to eat, thrive, reproduce, and avoid danger by patterns that range from the tropisms of single cells to instinct, choice, memory, learning, individuality, and creativity in more complex life-forms. The problem of when to "pull the plug" shows that we don't even know for sure how to diagnose death. Mechanistic chemistry isn't adequate to understand these enigmas of life, and it now acts as a barrier to studying them. Erwin Chargaff, the biochemist who discovered base pairing in DNA and thus opened the way for understanding gene structure, phrased our dilemma precisely when he wrote of biology, "No other science deals in its very name with a subject that it cannot define."

Given the present climate, I've been a lucky man. I haven't been a good, efficient doctor in the modern sense. I've spent far too much time on a few incurable patients whom no one else wanted, trying to find out how our ignorance failed them. I've been able to tack against the prevailing winds of orthodoxy and indulge my passion for experiment. In so doing I've been part of a little-known research effort that has made a new start toward a definition of life.

My research began with experiments on regeneration, the ability of some animals, notably the salamander, to grow perfect replacements for parts of the body that have been destroyed. These studies, described in Part 1, led to the discovery of a hitherto unknown aspect of animal life—the existence of electrical currents in parts of the nervous system. This breakthrough in turn led to a better understanding of bone fracture healing, new possibilities for cancer research, and the hope of human regeneration—even of the heart and spinal cord—in the not too distant future, advances that are discussed in Parts 2 and 3. Finally, a knowledge of life's electrical dimension has yielded fundamental insights (considered in Part 4) into pain, healing, growth, consciousness, the nature of life itself, and the dangers of our electromagnetic technology.

I believe these discoveries presage a revolution in biology and medicine. One day they may enable the physician to control and stimulate healing at will. I believe this new knowledge will also turn medicine in the direction of greater humility, for we should see that whatever we achieve pales before the self-healing power latent in all organisms. The results set forth in the following pages have convinced me that our understanding of life will always be imperfect. I hope this realization will make medicine no less a science, yet more of an art again. Only then can it deliver its promised freedom from disease.

Part 1

Growth and Regrowth

Salamander: energy's seed sleeping interred in the marrow . . .

—OCTAVIO PAZ

One

Hydra's Heads and Medusa's Blood

There is only one health, but diseases are many. Likewise, there appears to be one fundamental force that heals, although the myriad schools of medicine all have their favorite ways of cajoling it into action.

Our prevailing mythology denies the existence of any such generalized force in favor of thousands of little ones sitting on pharmacists' shelves, each one potent against only a few ailments or even a part of one. This system often works fairly well, especially for treatment of bacterial diseases, but it's no different in kind from earlier systems in which a specific saint or deity, presiding over a specific healing herb, had charge of each malady and each part of the body. Modern medicine didn't spring full-blown from the heads of Pasteur and Lister a hundred years ago.

If we go back further, we find that most medical systems have combined such specifics with a direct, unitary appeal to the same vital principle in all illnesses. The inner force can be tapped in many ways, but all are variations of four main, overlapping patterns: faith healing, magic healing, psychic healing, and spontaneous healing. Although science derides all four, they sometimes seem to work as well for degenerative diseases and long-term healing as most of what Western medicine can offer.

Faith healing creates a trance of belief in both patient and practitioner, as the latter acts as an intercessor or conduit between the sick mortal and a presumed higher power. Since failures are usually ascribed to a lack of faith by the patient, this brand of medicine has always been a gold mine for charlatans. When bona fide, it seems to be an escalation of the placebo effect, which produces improvement in roughly one third of subjects who *think* they're being treated but are actually being given dummy pills in tests of new drugs. Faith healing requires even more confidence from the patient, so the disbeliever probably *can* prevent a cure and settle for the poor satisfaction of "I told you so." If even a few of these oft-attested cases are genuine, however, the healed one suddenly finds faith turned into certainty as the withered arm aches with unaccustomed sensation, like a starved animal waking from hibernation.

Magical healing shifts the emphasis from the patient's faith to the doctor's trained will and occult learning. The legend of Teta, an Egyptian magician from the time of Khufu (Cheops), builder of the Great Pyramid, can serve as an example. At the age of 110, Teta was summoned into the royal presence to demonstrate his ability to rejoin a severed head to its body, restoring life. Khufu ordered a prisoner beheaded, but Teta discreetly suggested that he would like to confine himself to laboratory animals for the moment. So a goose was decapitated. Its body was laid at one end of the hall, its head at the other. Teta repeatedly pronounced his words of power, and each time, the head and body twitched a little closer to each other, until finally the two sides of the cut met. They quickly fused, and the bird stood up and began cackling. Some consider the legendary miracles of Jesus part of the same ancient tradition, learned during Christ's precocious childhood in Egypt.

Whether or not we believe in the literal truth of these particular accounts, over the years so many otherwise reliable witnesses have testified to healing "miracles" that it seems presumptuous to dismiss them all as fabrications. Based on the material presented in this book, I suggest Coleridge's "willing suspension of disbelief" until we understand healing better. Shamans apparently once served at least some of their patients well, and still do where they survive on the fringes of the industrial world. Magical medicine suggests that our search for the healing power isn't so much an exploration as an act of remembering something that was once intuitively ours, a form of recall in which the knowledge is passed on or awakened by initiation and apprenticeship to the man or woman of power.

Sometimes, however, the secret needn't be revealed to be used. Many psychic healers have been studied, especially in the Soviet Union, whose gift is unconscious, unsought, and usually discovered by accident. One who demonstrated his talents in the West was Oskar Estebany. A Hungarian Army colonel in the mid-1930s, Estebany noticed that horses he groomed got their wind back and recovered from illnesses faster than those cared for by others. He observed and used his powers informally for years, until, forced to emigrate after the 1956 Hungarian revolution, he settled in Canada and came to the attention of Dr. Bernard Grad, a biologist at McGill University. Grad found that Estebany could accelerate the healing of measured skin wounds made on the backs of mice, as compared with controls. He didn't let Estebany touch the animals, but only place his hands near their cages, because handling itself would have fostered healing. Estebany also speeded up the growth of barley plants and reactivated ultraviolet-damaged samples of the stomach enzyme trypsin in much the same way as a magnetic field, even though no magnetic field could be detected near his body with the instruments of that era.

The types of healing we've considered so far have trance and touch as common factors, but some modes don't even require a healer. The spontaneous miracles at Lourdes and other religious shrines require only a vision, fervent prayer, perhaps a momentary connection with a holy relic, and intense concentration on the diseased organ or limb. Other reports suggest that only the intense concentration is needed, the rest being aids to that end. When Diomedes, in the fifth book of the *Iliad,* dislocates Aeneas' hip with a rock, Apollo takes the Trojan hero to a temple of healing and restores full use of his leg within minutes. Hector later receives the same treatment after a rock in the chest fells him. We could dismiss these accounts as the hyperbole of a great poet if Homer weren't so realistic in other battlefield details, and if we didn't have similar accounts of soldiers in recent wars recovering from "mortal" wounds or fighting on while oblivious to injuries that would normally cause excruciating pain. British Army surgeon Lieutenant Colonel H. K. Beecher described 225 such cases in print after World War II. One soldier at Anzio in 1943, who'd had eight ribs severed near the spine by shrapnel, with punctures of the kidney and lung, who was turning blue and near death, kept trying to get off his litter because he thought he was lying on his rifle. His bleeding abated, his color returned, and the massive wound began to heal after no treatment but an insignificant dose of sodium amytal, a weak sedative given him because there was no morphine.

These occasional prodigies of battlefield stress strongly resemble the ability of yogis to control pain, stop bleeding, and speedily heal wounds with their will alone. Biofeedback research at the Menninger Foundation and elsewhere has shown that some of this same power can be tapped in people with no yogic training. That the will can be applied to the body's ills has also been shown by Norman Cousins in his resolute conquest by laugh therapy of ankylosing spondylitis, a crippling disease in which the spinal discs and ligaments solidify like bone, and by some similar successes by users of visualization techniques to focus the mind against cancer.

Unfortunately, no approach is a sure thing. In our ignorance, the common denominator of all healing—even the chemical cures we profess to understand—remains its mysteriousness. Its unpredictability has bedeviled doctors throughout history. Physicians can offer no reason why one patient will respond to a tiny dose of a medicine that has no effect on another patient in ten times the amount, or why some cancers go into remission while others grow relentlessly unto death.

By whatever means, if the energy is successfully focused, it results in a marvelous transformation. What seemed like an inexorable decline suddenly reverses itself. Healing can almost be *defined* as a miracle. Instant regrowth of damaged parts and invincibility against disease are commonplaces of the divine world. They continually appear even in myths that have nothing to do with the theme of healing itself. Dead Vikings went to a realm where they could savor the joys of killing all day long, knowing their wounds would heal in time for the next day's mayhem. Prometheus' endlessly regrowing liver was only a clever torture arranged by Zeus so that the eagle sent as punishment for the god's delivery of fire to mankind could feast on his most vital organ forever— although the tale also suggests that the prehistoric Greeks knew something of the liver's ability to enlarge in compensation for damage to it.

The Hydra was adept at these offhand wonders, too. This was the monster Hercules had to kill as his second chore for King Eurystheus. The beast had somewhere between seven and a hundred heads, and each time Hercules cut one off, two new ones sprouted in its place—until the hero got the idea of having his nephew Iolaus cauterize each neck as soon as the head hit the ground.

In the eighteenth century the Hydra's name was given to a tiny aquatic animal having seven to twelve "heads," or tentacles, on a hollow, stalklike body, because this creature can regenerate. The mythic Hydra remains a symbol of that ability, possessed to some degree by most animals, including us.

Generation, life's normal transformation from seed to adult, would seem as unearthly as regeneration if it were not so commonplace. We see the same kinds of changes in each. The Greek hero Cadmus grows an army by sowing the teeth of a dragon he has killed. The primeval serpent makes love to the World Egg, which hatches all the creatures of the earth. God makes Adam from Eve's rib, or vice versa in the later version. The Word of God commands life to unfold. The genetic words encoded in DNA spell out the unfolding. At successive but still limited levels of understanding, each of these beliefs tries to account for the beautifully bizarre metamorphosis. And if some savage told us of a magical worm that built a little windowless house, slept there a season, then one day emerged and flew away as a jeweled bird, we'd laugh at such superstition if we'd never seen a butterfly.

The healer's job has always been to release something not understood, to remove obstructions (demons, germs, despair) between the sick patient and the force of life driving obscurely toward wholeness. The means may be direct—the psychic methods mentioned above—or indirect: Herbs can be used to stimulate recovery; this tradition extends from prehistoric wisewomen through the Greek herbal of Dioscorides and those of Renaissance Europe, to the prevailing drug therapies of the present. Fasting, controlled nutrition, and regulation of living habits to avoid stress can be used to coax the latent healing force from the sick body; we can trace this approach back from today's naturopaths to Galen and Hippocrates. Attendants at the healing temples of ancient Greece and Egypt worked to foster a dream in the patient that would either start the curative process in sleep or tell what must be done on awakening. This method has gone out of style, but it must have worked fairly well, for the temples were filled with plaques inscribed by grateful patrons who'd recovered. In fact, this mode was so esteemed that Aesculapius, the legendary doctor who originated it, was said to have been given two vials filled with the blood of Medusa, the snaky-haired witch-queen killed by Perseus. Blood from her left side restored life, while that from her right took it away—and that's as succinct a description of the tricky art of medicine as we're likely to find.

The more I consider the origins of medicine, the more I'm convinced that all true physicians seek the same thing. The gulf between folk therapy and our own stainless-steel version is illusory. Western medicine springs from the same roots and, in the final analysis, acts through the same little-understood forces as its country cousins. Our doctors ignore this kinship at their—and worse, their patients'—peril. All worthwhile medical research and every medicine man's intuition is part of the same quest for knowledge of the same elusive healing energy.

Failed Healing in Bone

As an orthopedic surgeon, I often pondered one particular breakdown of that energy, my specialty's major unsolved problem—nonunion of fractures. Normally a broken bone will begin to grow together in a few weeks if the ends are held close to each other without movement. Occasionally, however, a bone will refuse to knit despite a year or more of casts and surgery. This is a disaster for the patient and a bitter defeat for the doctor, who must amputate the arm or leg and fit a prosthetic substitute.

Throughout this century, most biologists have been sure only chemical processes were involved in growth and healing. As a result, most work on nonunions has concentrated on calcium metabolism and hormone relationships. Surgeons have also "freshened," or scraped, the fracture surfaces and devised ever more complicated plates and screws to hold the bone ends rigidly in place. These approaches seemed superficial to me. I doubted that we would ever understand the failure to heal unless we truly understood healing itself.

When I began my research career in 1958, we already knew a lot about the logistics of bone mending. It seemed to involve two separate processes, one of which looked altogether different from healing elsewhere in the human body. But we lacked any idea of what set these processes in motion and controlled them to produce a bone bridge across the break.

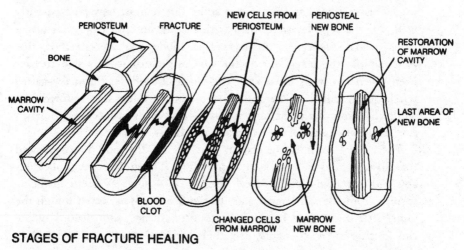

STAGES OF FRACTURE HEALING

Every bone is wrapped in a layer of tough, fibrous collagen, a protein that's a major ingredient of bone itself and also forms the connective tissue or "glue" that fastens all our cells to each other. Underneath the collagen envelope are the cells that produce it, right next to the bone; together the two layers form the periosteum. When a bone breaks, these periosteal cells divide in a particular way. One of the daughter cells stays where it is, while the other one migrates into the blood clot surrounding the fracture and changes into a closely related type, an osteoblast, or bone-forming cell. These new osteoblasts build a swollen ring of bone, called a callus, around the break.

Another repair operation is going on inside the bone, in its hollow center, the medullary cavity. In childhood the marrow in this cavity actively produces red and white blood cells, while in adulthood most of the marrow turns to fat. Some active marrow cells remain, however, in the porous convolutions of the inner surface. Around the break a new tissue forms from the marrow cells, most readily in children and young animals. This new tissue is unspecialized, and the marrow cells seem to form it not by increasing their rate of division, as in the callus-forming periosteal cells, but by reverting to a primitive, neo-embryonic state. The unspecialized former marrow cells then change into a type of primitive cartilage cells, then into mature cartilage cells, and finally into new bone cells to help heal the break from inside. Under a microscope, the changes seen in cells from this internal healing area, especially from children a week or two after the bone was broken, seem incredibly chaotic, and they look frighteningly similar to highly malignant bone-cancer cells. But in most cases their transformations are under control, and the bone heals.

Dr. Marshall Urist, one of the great researchers in orthopedics, was to conclude in the early 1960s that this second type of bone healing is an evolutionary throwback, the only kind of regeneration that humans share with all other vertebrates. Regeneration in this sense means the regrowth of a complex body part, consisting of several different kinds of cells, in a fashion resembling the original growth of the same part in the embryo, in which the necessary cells differentiate from simpler cells or even from seemingly unrelated types. This process, which I'll call true regeneration, must be distinguished from two other forms of healing. One, sometimes considered a variety of regeneration, is physiological repair, in which small wounds and everyday wear within a single tissue are made good by nearby cells of the same type, which merely proliferate to close the gap. The other kind of healing occurs when a wound is too big for single-tissue repair but the animal lacks the true regenerative competence to restore the damaged part. In this case the injury is simply patched over as well as possible with collagen fibers, forming a scar. Since true regeneration is most closely related to embryonic development and is generally strongest in simple animals, it may be considered the most fundamental mode of healing.

Nonunions failed to knit, I reasoned, because they were missing something that triggered and controlled normal healing. I'd already begun to wonder if the inner area of bone mending might be a vestige of true regeneration. If so, it would likely show the control process in a clearer or more basic form than the other two kinds of healing. I figured I stood little chance of isolating a clue to it in the multilevel turmoil of a broken bone itself, so I resolved to study regeneration alone, as it occurred in other animals.

A Fable Made Fact

Regeneration happens all the time in the plant kingdom. Certainly this knowledge was acquired very early in mankind's history. Besides locking up their future generations in the mysterious seed, many plants, such as grapevines, could form a new plant from a single part of the old. Some classical authors had an inkling of animal regeneration—Aristotle mentions that the eyes of very young swallows recover from injury, and Pliny notes that lost "tails" of octopi and lizards regrow. However, regrowth was thought to be almost exclusively a plant prerogative.

The great French scientist René Antoine Ferchault de Réaumur made the first scientific description of animal regeneration in 1712. Réaumur devoted all his life to the study of "insects," which at that time meant all invertebrates, everything that was obviously "lower" than lizards, frogs, and fish. In studies of crayfish, lobsters, and crabs, Réaumur proved the claims of Breton fishermen that these animals could regrow lost legs. He kept crayfish in the live-bait well of a fishing boat, removing a claw from each and observing that the amputated extremity reappeared in full anatomical detail. A tiny replica of the limb took shape inside the shell; when the shell was discarded at the next molting season, the new limb unfolded and grew to full size.

Réaumur was one of the scientific geniuses of his time. Elected to the Royal Academy of Sciences when only twenty-four, he went on to invent tinned steel, Réaumur porcelain (an opaque white glass), imitation pearls, better ways of forging iron, egg incubators, and the Réaumur thermometer, which is still used in France. At the age of sixty-nine he isolated gastric juice from the stomach and described its digestive function. Despite his other accomplishments, "insects" were his life's love (he never married), and he probably was the first to conceive of the vast, diverse population of life-forms that this term encompassed. He rediscovered the ancient royal purple dye from *Murex trunculus* (a marine mollusk), and his work on spinning a fragile, filmy silk from spider webs was translated into Manchu for the Chinese emperor. He was the first to elucidate the social life and sexually divided caste system of bees. Due to his eclipse in later years by court-supported scientists who valued "common sense" over observation, Réaumur's exhaustive study of ants wasn't published until 1926. In the interim it had taken several generations of formicologists to cover the same ground, including the description of winged ants copulating in flight and proof that they aren't a separate species but the sexual form of wingless ants. In 1734 he published the first of six volumes of his *Natural History of Insects,* a milestone in biology.

Réaumur made so many contributions to science that his study of regeneration was overlooked for decades. At that time no one really cared what strange things these unimportant animals did. However, all of the master's work was well known to a younger naturalist, Abraham Trembley of Geneva, who supported himself, as did many educated men of that time, by serving as a private tutor for sons of wealthy families. In 1740, while so employed at an estate near The Hague, in Holland, Trembley was examining with a hand lens the small animals living in freshwater ditches and ponds. Many had been described by Réaumur, but Trembley chanced upon an odd new one. It was no more than a quarter of an inch long and faintly resembled a squid, having a cylindrical body topped with a crown of tentacles. However, it was a startling green color. To Trembley, green meant vegetation, but if this was a plant, it was a mighty peculiar one. When Trembley agitated the water in its dish, the tentacles contracted and the body shrank down to a nubbin, only to reexpand after a period of quiet. Strangest of all, he saw that the creature "walked" by somersaulting end over end.

Since they had the power of locomotion, Trembley would have assumed that these creatures were animals and moved on to other observations, if he hadn't chanced to find a species colored green by symbiotic algae. To settle the animal-plant question, he decided to cut some in half. If they regrew, they must be plants with the unusual ability to walk, while if they couldn't regenerate, they must be green animals.

Trembley soon entered into a world that exceeded his wildest dreams. He divided the polyps, as he first called them, in the middle of their stalks. He then had two short pieces of stalk, one with attached tentacles, each of which contracted down to a tiny dot. Patiently watching, Trembley saw the two pieces later expand. The tentacle portion began to move normally, as though it were a complete organism. The other portion lay inert and apparently dead. Something must have made Trembley continue the experiment, for he watched this motionless object for nine days, during which nothing happened. He then noted that the cut

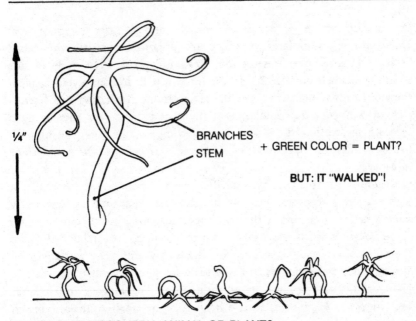

1/4"

BRANCHES

STEM

+ GREEN COLOR = PLANT?

BUT: IT "WALKED"!

TREMBLEY'S DISCOVERY: ANIMAL OR PLANT?

end had sprouted three little "horns," and within a few more days the complete crown of tentacles had been restored. Trembley now had two complete polyps as a result of cutting one in half! However, even though they regenerated, more observations convinced Trembley that the creatures were really animals. Not only did they move and walk, but their arms captured tiny water fleas and moved them to the "mouth," located in the center of the ring of tentacles, which promptly swallowed the prey.

Trembley, then only thirty-one, decided to make sure he was right by having the great Réaumur confirm his findings before he published them and possibly made a fool of himself. He sent specimens and detailed notes to Réaumur, who confirmed that this was an animal with amazing powers of regeneration. Then he immediately read Trembley's letters and showed his specimens to an astounded Royal Academy early in 1741. The official report called Trembley's polyp more marvelous than the phoenix or the mythical serpent that could join together after being cut in two, for these legendary animals could only reconstitute themselves, while the polyp could make a duplicate. Years later Réaumur was still thunderstruck. As he wrote in Volume 6 of his series on insects, "This is a fact that I cannot accustom myself to seeing, after having seen and re-seen it hundreds of times."

This was just the beginning, however. Trembley's polyps performed even more wondrous feats. When cut lengthwise, each half of the stalk healed over without a scar and proceeded to regrow the missing tentacles. Trembley minced some polyps into as many pieces as he could manage, finding that a complete animal would regrow from each piece, as long as it included a remnant of the central stalk. In one instance he quartered one of the creatures, then cut each resulting polyp into three or four pieces, until he had made fifty animals from one.

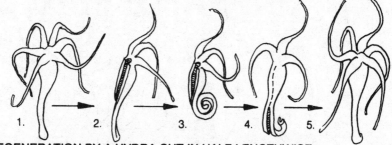

REGENERATION BY A HYDRA CUT IN HALF LENGTHWISE

His most famous experiment was the one that led him to name his polyp "hydra." He found that by splitting the head lengthwise, leaving the stalk intact, he could produce one animal with two crowns of tentacles. By continuing the process he was able to get one animal with seven heads. When Trembley lopped them off, each one regrew, just like the mythical beast's. But nature went legend one better: Each severed head went on to form a complete new animal as well.

Such experiments provided our first proof that entire animals can regenerate, and Trembley went on to observe that hydras could reproduce by simple budding, a small animal appearing on the side of the stalk and growing to full size. The implications of these discoveries were so revolutionary that Trembley delayed publishing a full account of his work until he'd been prodded by Réaumur and preceded in print by several others. The sharp division between plant and animal suddenly grew blurred, suggesting a common origin with some kind of evolution; basic assumptions about life had to be rethought. As a result, Trembley's observations weren't enthusiastically embraced by all. They inflamed several old arguments and offended many of the old guard. In this respect Trembley's mentor Réaumur was a most unusual scientist for his time, and indeed for all time. Despite his prominence, he was ready to espouse radically new ideas and, most important, he didn't steal the ideas of others, an all too common failing among scientists.

A furious debate was raging at the time of Trembley's announcement. It concerned the origin of the individual—how the chicken came from

A HYDRA REGROWS ITS MOUTH AND TENTACLES

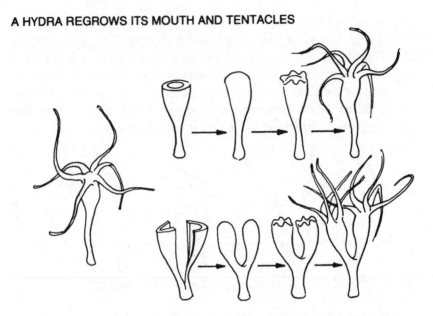

A PARTIALLY SPLIT HYDRA'S BODY GROWS TWO "HEADS"

the egg, for example. When scientists examined the newly laid egg, there wasn't much there except two liquids, the white and the yolk, neither of which had any discernible structure, let alone anything resembling a chicken.

There were two opposite theories. The older one, derived from Aristotle, held that each animal in all its complexity developed from simple organic matter by a process called epigenesis, akin to our modern concept of cell differentiation. Unfortunately, Trembley himself was the first person to witness cell division under the microscope, and he didn't realize that it was the normal process by which all cells multiplied. In an era knowing nothing of genes and so little of cells, yet beginning to insist on logical, scientific explanations, epigenesis seemed more and more absurd. What could possibly transform the gelatin of eggs and sperm into a frog or a human, without invoking that tired old deus ex machina the spirit, or inexplicable spark of life—*unless* the frog or person already existed in miniature inside the generative slime and merely grew in the course of development?

The latter idea, called preformation, had been ascendant for at least fifty years. It was so widely accepted that when the early microscopists studied drops of semen, they dutifully reported a little man, called a homunculus, encased in the head of each sperm—a fine example of science's capacity for self-delusion. Even Réaumur, when he failed to find tiny butterfly wings inside caterpillars, assumed they were there but were too small to be seen. Only a few months before Trembley began slicing hydras, his cousin, Genevan naturalist Charles Bonnet, had proven (in an experiment suggested by the omnipresent Réaumur) that female aphids usually reproduced parthenogenetically (without mating). To Bonnet this demonstrated that the tiny adult resided in the egg, and he became the leader of the ovist preformationists.

The hydra's regeneration, and similar powers in starfish, sea anemones, and worms, put the scientific establishment on the defensive. Réaumur had long ago realized that preformation couldn't explain how a baby inherited traits from both father and mother. The notion of two homunculi fusing into one seed seemed farfetched. His regrowing crayfish claws showed that each leg would have to contain little preformed legs scattered throughout. And since a regenerated leg could be lost and replaced many times, the proto-legs would have to be very numerous, yet no one had ever found any.

Regeneration therefore suggested some form of epigenesis—perhaps without a soul, however, for the hydra's anima, if it existed, was divisible along with the body and indistinguishable from it. It seemed as though some forms of matter itself possessed the spark of life. For lack of knowledge of cells, let alone chromosomes and genes, the epigeneticists were unable to prove their case. Each side could only point out the other's inconsistencies, and politics gave preformationism the edge.

No wonder nonscientists often grew impatient of the whole argument. Oliver Goldsmith and Tobias Smollett mocked the naturalists for missing nature's grandeur in their myopic fascination with "muck-flies." Henry Fielding lampooned the discussion in a skit about the regeneration of money. Diderot thought of hydras as composite animals, like swarms of bees, in which each particle had a vital spark of its own, and lightheartedly suggested there might be "human polyps" on Jupiter and Saturn. Voltaire was derisively skeptical of attempts to infer the nature of the soul, animal or human, from these experiments. Referring in 1768 to the regenerating heads of snails, he asked, "What happens to its sensorium, its memory, its store of ideas, its soul, when its head has been cut off? How does all this come back? A soul that is reborn is an extremely curious phenomenon." Profoundly disturbed by the whole affair, for a long time he simply refused to believe in animal regeneration, calling the hydra "a kind of small rush."

It was no longer possible to doubt the discovery after the work of Lazzaro Spallanzani, an Italian priest for whom science was a full-time hobby. In a career spanning the second half of the eighteenth century, Spallanzani discovered the reversal of plant transpiration between light and darkness, and advanced our knowledge of digestion, volcanoes, blood circulation, and the senses of bats, but his most important work concerned regrowth. In twenty years of meticulous observation, he studied regeneration in worms, slugs, snails, salamanders, and tadpoles. He set new standards for thoroughness, often dissecting the amputated parts to make sure he'd removed them whole, then dissecting the replacements a few months later to confirm that all the parts had been restored.

Spallanzani's most important contribution to science was his discovery of the regenerative abilities of the salamander. It could replace its tail and limbs, all at once if need be. A young one performed this feat for Spallanzani six times in three months. He later found that the salamander could also replace its jaw and the lenses of its eyes, and then went on to establish two general rules of regeneration: Simple animals can regenerate more fully than complex ones, or, in modern terms, the ability to regenerate declines as one moves up the evolutionary scale. (The salamander is the main exception.) In ontogenetic parallel, if a species can regenerate, younger individuals do it better than older ones.

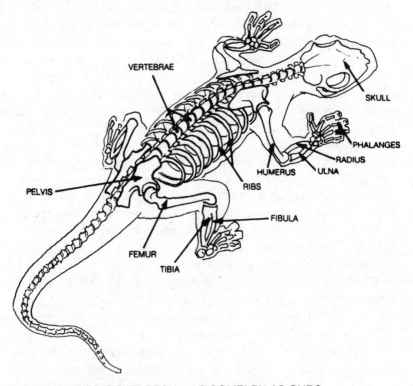

THE SALAMANDER'S SKELETON—AS COMPLEX AS OURS

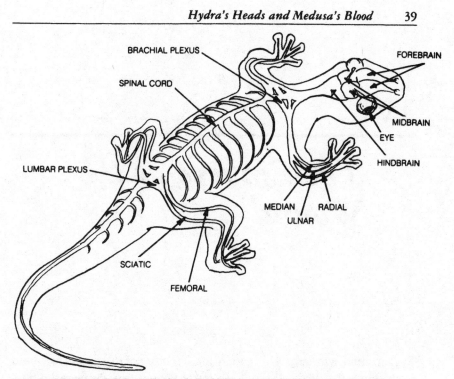

BRACHIAL PLEXUS

SPINAL CORD

FOREBRAIN

MIDBRAIN

EYE

HINDBRAIN

LUMBAR PLEXUS

MEDIAN   RADIAL

ULNAR

SCIATIC

FEMORAL

THE SALAMANDER'S NERVOUS SYSTEM—FORERUNNER OF OURS

This early regeneration research, Spallanzani's in particular, was a benchmark in modern biology. Gentlemanly observations buttressed by "common sense" gave way to a more rigorous kind of examination in which *nothing* was taken for granted. It had been "known" for perhaps ten thousand years that plants could regenerate and animals couldn't. To many zoologists, even twenty years after Trembley's initial discovery, the few known exceptions only proved the rule, for octopi, crayfish, hydras, worms, and snails seemed so unlike humans or the familiar mammals that they hardly counted. The lizard, the only other vertebrate regenerator then known, could manage no more than an imperfect tail. But the salamander—here was an animal we could relate to! This was no worm or snail or microscopic dot, but a four-limbed, two-eyed vertebrate that could walk and swim. While its legendary ability to withstand fire had been disproven, its body was big enough and its anatomy similar enough to ours to be taken seriously. Scientists could no longer assume that the underlying process had nothing to do with us. In fact, the questions with which Spallanzani ended his first report on the salamander have haunted biologists ever since: "Is it to be hoped that [higher animals] may acquire [the same power] by some useful dispositions? and should the flattering expectation of obtaining this advantage for ourselves be considered entirely as chimerical?"

TWO

The Embryo at the Wound

Regeneration was largely forgotten for a century. Spallanzani had been so thorough that little else could be learned about it with the techniques of the time. Moreover, although his work strongly supported epigenesis, its impact was lost because the whole debate was swallowed up in the much larger philosophical conflict between vitalism and mechanism. Since biology includes the study of our own essence, it's the most emotional science, and it has been the battleground for these two points of view throughout its history. Briefly, the vitalists believed in a spirit, called the anima or *élan vital,* that made living things fundamentally different from other substances. The mechanists believed that life could ultimately be understood in terms of the same physical and chemical laws that governed nonliving matter, and that only ignorance of these forces led people to invoke such hokum as a spirit. We'll take up these issues in more detail later, but for now we need only note that the vitalists favored epigenesis, viewed as an imposition of order on the chaos of the egg by some intangible "vital" force. The mechanists favored preformation. Since science insisted increasingly on material explanations for everything, epigenesis lost out despite the evidence of regeneration.

Mechanism dominated biology more and more, but some problems remained. The main one was the absence of the little man in the sperm. Advances in the power and resolution of microscopes had clearly shown that no one was there. Biologists were faced with the generative slime again, featureless goo from which, slowly and magically, an organism appeared.

After 1850, biology began to break up into various specialties. Embryology, the study of development, was named and promoted by Darwin himself, who hoped (in vain) that it would reveal a precise history of evolution (phylogeny) recapitulated in the growth process (ontogeny). In the 1880s, embryology matured as an experimental science under the leadership of two Germans, Wilhelm Roux and August Weismann. Roux studied the stages of embryonic growth in a very restricted, mechanistic way that revealed itself even in the formal Germanic title, *Entwicklungsmechanik* ("developmental mechanics"), that he applied to the whole field. Weismann, however, was more interested in how inheritance passed the instructions for embryonic form from one generation to the next. One phenomenon—mitosis, or cell division—was basic to both transactions. No matter how embryos grew and hereditary traits were transferred, both processes had to be accomplished by cellular actions.

Although we're taught in high school that Robert Hooke discovered the cell in 1665, he really discovered that cork was full of microscopic holes, which he called cells because they looked like little rooms. The idea that they were the basic structural units of all living things came from Theodor Schwann, who proposed this cell theory in 1838. However, even at that late date, he didn't have a clear idea of the origin of cells. Mitosis was unknown to him, and he wasn't too sure of the distinction between plants and animals. His theory wasn't fully accepted until two other German biologists, F. A. Schneider and Otto Bütschli, reintroduced Schwann's concept and described mitosis in 1873.

Observations of embryogenesis soon confirmed its cellular basis. The fertilized egg was exactly that, a seemingly unstructured single cell. Embryonic growth occurred when the fertilized egg divided into two other cells, which promptly divided again. Their progeny then divided, and so on. As they proliferated, the cells also differentiated; that is, they began to show specific characteristics of muscle, cartilage, nerve, and so forth. The creature that resulted obviously had several increasingly complex levels of organization; however, Roux and Weismann had no alternative but to concentrate on the lowest one, the cell, and try to imagine how the inherited material worked at that level.

Weismann proposed a theory of "determiners," specific chemical structures coded for each cell type. The fertilized egg contained all the determiners, both in type and in number, needed to produce every cell in the body. As cell division proceeded, the daughter cells each received half of the previous stock of determiners, until in the adult each cell possessed only one. Muscle cells contained only the muscle determiner, nerve cells only the one for nerves, and so on. This meant that *once a cell's function had been fixed, it could never be anything but that one kind of cell.*

In one of his first experiments, published in 1888, Roux obtained powerful support for this concept. He took fertilized frog eggs, which were large and easy to observe, and waited until the first cell division had occurred. He then separated the two cells of this incipient embryo. According to the theory, each cell contained enough determiners to make half an embryo, and that was exactly what Roux got—two half-embryos. It was hard to argue with such a clear-cut result, and the determiner theory was widely accepted. Its triumph was a climactic victory for the mechanistic concept of life, as well.

One of vitalism's last gasps came from the work of another German embryologist, Hans Driesch. Initially a firm believer in *Entwicklungs-mechanik,* Driesch later found its concepts deficient in the face of life's continued mysteries. For example, using sea urchin eggs, he repeated Roux's famous experiment and obtained a *whole* organism instead of a half. Many other experiments convinced Driesch that life had some special innate drive, a process that went against known physical laws. Drawing on the ancient Greek idea of the anima, he proposed a nonmaterial, vital factor that he called entelechy. The beginning of the twentieth century wasn't a propitious time for such an idea, however, and it wasn't popular.

Mechanics of Growth

As the nineteenth century drew to a close and the embryologists continued to struggle with the problems of inheritance, they found they still needed a substitute for the homunculus. Weismann's determiners worked fine for embryonic growth, but regeneration was a glaring exception, and one that didn't prove the rule. The original theory had no provision for a limited replay of growth to replace a part lost after development was finished. Oddly enough, the solution had already been provided by a man almost totally forgotten today, Theodor Heinrich Boveri.

Working at the University of Munich in the 1880s, Boveri discovered almost every detail of cell division, including the chromosomes. Not until the invention of the electron microscope did anyone add materially to his original descriptions. Boveri found that all nonsexual cells of any one species contained the same number of chromosomes. As growth proceeded by mitosis, these chromosomes split lengthwise to make two of each so that each daughter cell then had the same number of chromosomes. The egg and sperm, dividing by a special process called meiosis, wound up with exactly half that number, so that the fertilized egg would start out with a full complement, half from the father and half from the mother. He reached the obvious conclusions that the chromosomes transmitted heredity, and that each one could exchange smaller units of itself with its counterpart from the other parent.

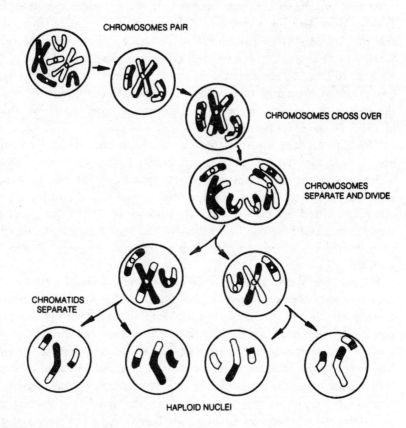

MEIOSIS—FORMATION OF SEX CELLS

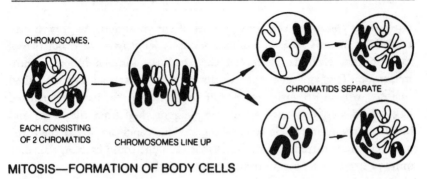

CHROMOSOMES,

EACH CONSISTING
OF 2 CHROMATIDS

CHROMOSOMES LINE UP

CHROMATIDS SEPARATE

MITOSIS—FORMATION OF BODY CELLS

At first this idea wasn't well received. It was strenuously opposed by Thomas Hunt Morgan, a respected embryologist at Columbia University and the first American participant in this saga. Later, when Morgan found that the results of his own experiments agreed with Boveri's, he went on to describe chromosome structure in more detail, charting specific positions, which he called genes, for inherited characteristics. Thus the science of genetics was born, and Morgan received the Nobel Prize in 1933. So much for Boveri.

Although Morgan was most famous for his genetics research on fruit flies, he got his start by studying salamander limb regeneration, about which he made a crucial observation. He found that the new limb was preceded by a mass of cells that appeared on the stump and resembled the unspecialized cell mass of the early embryo. He called this structure the blastema and later concluded that the problem of how a regenerated limb formed was identical to the problem of how an embryo developed from the egg.

Morgan postulated that the chromosomes and genes contained not only the inheritable characteristics but also the code for cell differentiation. A muscle cell, for example, would be formed when the group of genes specifying muscle were in action. This insight led directly to our modern understanding of the process: In the earliest stages of the embryo, every gene on every chromosome is active and available to every cell. As the organism develops, the cells form three rudimentary tissue layers—the endoderm, which develops into the glands and viscera; the mesoderm, which becomes the muscles, bones, and circulatory system; and the ectoderm, which gives rise to the skin, sense organs, and nervous system. Some of the genes are already being turned off, or repressed, at this stage. As the cells differentiate into mature tissues, only one specific set of genes stays switched on in each kind. Each set can make only certain types of messenger ribonucleic acid (mRNA), the "executive secretary" chemical by means of which DNA "instructs" the ribosomes (the cell's protein-factory organelles) to make the particular proteins that distinguish a nerve cell, for example, from a muscle or cartilage cell.

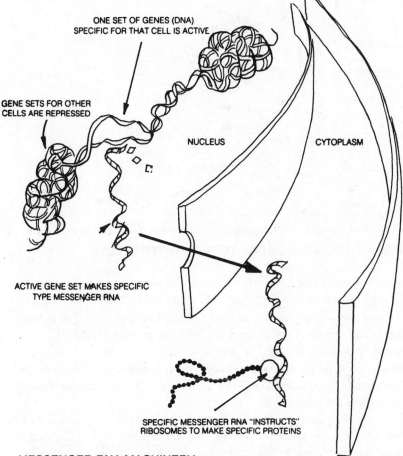

ONE SET OF GENES (DNA)
SPECIFIC FOR THAT CELL IS ACTIVE

GENE SETS FOR OTHER
CELLS ARE REPRESSED

NUCLEUS   CYTOPLASM

ACTIVE GENE SET MAKES SPECIFIC
TYPE MESSENGER RNA

SPECIFIC MESSENGER RNA "INSTRUCTS"
RIBOSOMES TO MAKE SPECIFIC PROTEINS

MESSENGER RNA MACHINERY

There's a superficial similarity between this genetic mechanism and the old determiner theory. The crucial difference is that, instead of determiners being segregated until only one *remains* in each cell, the genes are repressed until only one set remains *active* in each cell. However, *the entire genetic blueprint is carried by every cell nucleus.*

Science is a bit like the ancient Egyptian religion, which never threw old gods away but only tacked them onto newer deities until a bizarre hodgepodge developed. For some strange reason, science is equally reluctant to discard worn-out theories, and, even though there was absolutely no evidence to support it, one of Weismann's ideas was swallowed whole

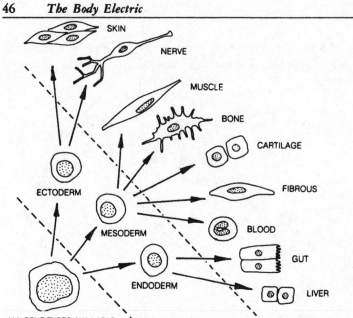

ALL GENE TYPES AVAILABLE   GENE REPRESSION BEGINNING   ONLY 1 GENE TYPE AVAILABLE

DIFFERENTIATION—FROM EGG TO SPECIALIZED CELLS

by the new science of genetics. This was the notion that differentiation was still a "one-way street," that cells could never *de*differentiate, that is, retrace their steps from a mature, specialized state to a primitive, unspecialized form. This assumption was made despite the fact that chromosomes now provided a plausible means for the reversal. Remember, *all* cells of the adult (except the egg and sperm) contain the full array of chromosomes. All the genes are still there, even though most of them are repressed.

It seems logical that what has been locked might also be unlocked when new cells are needed, but this idea was fought with unbelievable ferocity by the scientific establishment. It's difficult now to see why, since no principle of real importance was involved, except possibly a bit of the supremacy of the mechanistic outlook itself. The mechanists greeted the discovery of genes and chromosomes joyfully. Here at last was a replacement for the little man in the sperm! Perhaps it seemed that admitting dedifferentiation would have given life too much control over its own functions. Perhaps, once genes were considered the sole mechanism of life, they had to work in a nice, simple, mechanical way. As we shall see, this dogma created terrible difficulties for the study of regeneration.

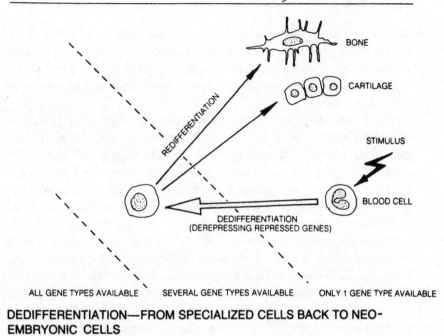

ALL GENE TYPES AVAILABLE   SEVERAL GENE TYPES AVAILABLE   ONLY 1 GENE TYPE AVAILABLE

DEDIFFERENTIATION—FROM SPECIALIZED CELLS BACK TO NEO-EMBRYONIC CELLS

Control Problems

After Morgan's work on salamander limb regrowth early in this century, hundreds of other experimenters studied the miracle again and again in many kinds of animals. Their labors revealed a number of general principles, such as:

- *Polarity.* A creature's normal relationships of front to back and top to bottom are preserved in the regenerate.
- *Gradients.* Regenerative ability is strongest in one area of an animal's body, gradually diminishing in all directions.
- *Dominance.* Some one particular section of the lost part is replaced first, followed by the others in a fixed sequence.
- *Induction.* Some parts actively trigger the formation of others later in the sequence.
- *Inhibition.* The presence of any particular part prevents the formation of a duplicate of itself or of other parts that come before that part in the sequence.

All the experiments led to one unifying conclusion: The overall structure, the shape, the pattern, of any animal is as real a part of its body as are its cells, heart, limbs, or teeth. Living things are called organisms because of the overriding importance of organization, and each part of the pattern somehow contains the information as to *what it is* in relation to the whole. The ability of this pattern to maintain itself reaches its height in the newts, mud puppies, and other amphibians collectively called salamanders.

The salamander, directly descended from the evolutionary prototype of all land vertebrates, is a marvelously complex animal, almost as complicated as a human. Its forelimb is basically the same as ours. Yet all its interrelated parts grow back in the proper order—the same interlocking bones and muscles, all the delicate wrist bones, the coordinated fingers—and they're wired together with the proper nerve and blood vessel connections.

The same day the limb is cut off, debris from dead cells is carried away in the bloodstream. Then some of the intact tissue begins to die back a short distance from the wound. During the first two or three days, cells of the epidermis—the outer layer of skin—begin to proliferate and migrate inward, covering the wound surface. The epidermis then thickens over the apex of the stump into a transparent tissue called the apical cap. This stage is finished in about a week.

By then, the blastema, the little ball of undifferentiated cells described by Morgan, has started to appear beneath the apical cap. This is the "organ" of regeneration, forming on the wound like a miniature embryo and very similar to the embryonic limb bud that gave rise to the leg in the first place. Its cells are totipotent, able to develop into all the different kinds of cells needed to reconstitute the limb.

The blastema is ready in about two weeks. Even as it's forming, the cells at its outer edge start dividing rapidly, changing the blastema's shape to a cone and providing a steady source of raw material—new cells—for growth. After about three weeks, the blastema cells at the inner edge begin to differentiate into specialized types and arrange themselves into tissues, beginning with a cartilage collar around the old bone shaft. Other tissues then form, and the new limb—beginning with a characteristic paddle shape that will become the hand—appears as though out of a mist. The elbow and long parts of the limb coalesce behind the hand, and the regrowth is complete (except for some slight enlargement) when the four digits reappear after about eight weeks.

This process, exquisitely beautiful and seemingly simple, is full of problems for biology. What organizes the growth? What is the control factor? How does the blastema "know" that it must make a foreleg instead of a hind leg? (The salamander never makes a mistake.) How does all the information about the missing parts get to these undifferentiated

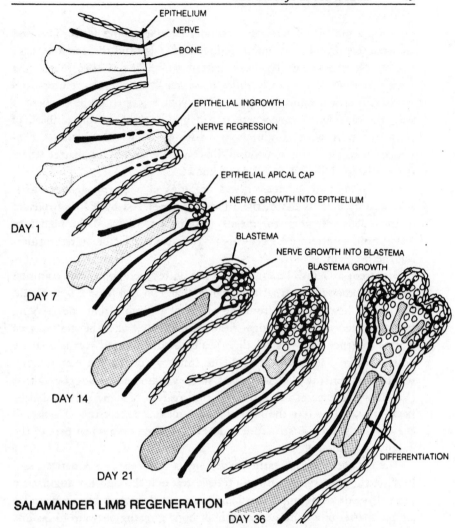

EPITHELIUM

NERVE

BONE

EPITHELIAL INGROWTH

NERVE REGRESSION

EPITHELIAL APICAL CAP

NERVE GROWTH INTO EPITHELIUM

DAY 1

BLASTEMA

NERVE GROWTH INTO BLASTEMA

BLASTEMA GROWTH

DAY 7

DAY 14

DIFFERENTIATION

DAY 21

SALAMANDER LIMB REGENERATION

DAY 36

cells, telling them what to become, which genes to activate, what proteins to make, where to position themselves? It's as if a pile of bricks were to spontaneously rearrange itself into a building, becoming not only walls but windows, light sockets, steel beams, and furniture in the process.

Answers were sought by transplanting the blastema to other positions on the animal. The experiments only made matters worse. If the blastema was moved within five to seven days after it first appeared, and grafted near the hind leg, it grew into a second hind leg, even though it came from an amputated foreleg. Well, that was okay. The body could be divided into "spheres of influence" or "organizational territories," each of which contained information on the local anatomy. A blastema put into a hind-limb territory naturally became a hind limb. This was an attractive theory, but unfounded. Exactly what did this territory consist of? No one knew. To make matters worse, it was then found that transplantation of a slightly older blastema from a foreleg stump to a hind-limb area produced a *foreleg*. The young blastema knew where it was; the older one knew where it had been! Somehow this pinhead of primitive cells with absolutely no distinguishing characteristics contained enough information to build a complete foreleg, no matter where it was placed. How? We still don't know.

One attempt at an answer was the idea of a morphogenetic field, advanced by Paul Weiss in the 1930s and developed by H. V. Brønsted in the 1950s. *Morphogenesis* means "origin of form," and the field idea was simply an attempt to get closer to the control factor by reformulating the problem.

Brønsted, a Danish biologist working on regeneration in the common flatworms known as planarians, found that two complete heads would form when he cut a strip from the center of a worm's front end, leaving two side pieces of the original head. Conversely, when he grafted two worms together side by side, their heads fused. Brønsted saw an analogy with a match flame, which could be split by cutting the match, then rejoined by putting the two halves side by side, and he suggested that part of the essence of life might be the creation of some such flamelike field. It would be like the field around a magnet except that it reflected the magnet's internal structure and held its shape even when part of the magnet was missing.

The idea grew out of earlier experiments by Weiss, an American embryologist, who stymied much creative research through his dogmatism yet still made some important contributions. Regrowth clearly wasn't a simple matter of a truncated muscle or bone growing outward to resume its original shape. Structures that were missing entirely—the hand, wrist, and bones of the salamander's lower forelimb, for example—also reappeared. Weiss found that redundant parts could be inserted, but the essential ones couldn't easily be eliminated. If an extra bone was implanted in the limb and the cut made through the two, the regenerate contained both. However, if a bone was completely removed and the incision allowed to heal, and the limb was then amputated through what would have been the middle of the missing bone, the regenerate produced that bone's lower half, like a ghost regaining its substance. Weiss suggested that other tissues besides bone could somehow project a field that included the arrangement of the bones. As a later student of regeneration, Richard Goss of Brown University, observed, "Apparently

THE LATE BLASTEMA ALREADY HAS ITS INSTRUCTIONS

THE EARLY BLASTEMA GETS INSTRUCTIONS FROM NEARBY TISSUES

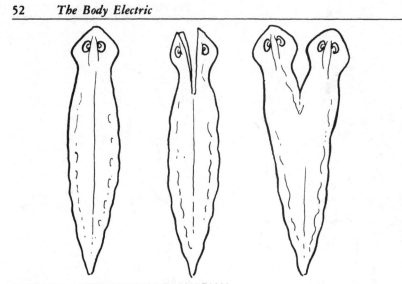

FORMING A TWO-HEADED PLANARIAN

each tissue of the stump can vote to be represented in the blastema, and some of them can even cast absentee ballots."

Any such field must be able to stimulate cells to switch various genes on and off, that is, to change their specialization. A large body of research on embryonic development has identified various chemical inducers, compounds that stimulate neighboring cells to differentiate in a certain fashion, producing the next type of cells needed. But these substances act only on the basis of simple diffusion; nothing in the way they operate can account for the way the process is controlled to express the overall pattern.

Another classic experiment helps clarify the problem. A salamander's hand can be amputated and the wrist stump sewn to its body. The wrist grows into the body, and nerves and blood vessels link up through the new connection. The limb now makes a U shape, connected to the body at both ends. It's then amputated at the shoulder to make a reversed limb, attached to the body at the wrist and ending with a shoulder joint. The limb then regenerates as though it had simply been cut off at the shoulder. The resulting limb looks like this: from the body sprouts the original wrist, forearm, elbow, upper arm, and shoulder, followed by a new upper arm, elbow, forearm, wrist, and hand. Why doesn't the regenerate conform to the sequence already established in this limb instead of following as closely as possible the body's pattern as a whole? Again, what is the control factor?

Information, and a monumental amount of it, is clearly passed from the body to the blastema. Our best method of information processing at

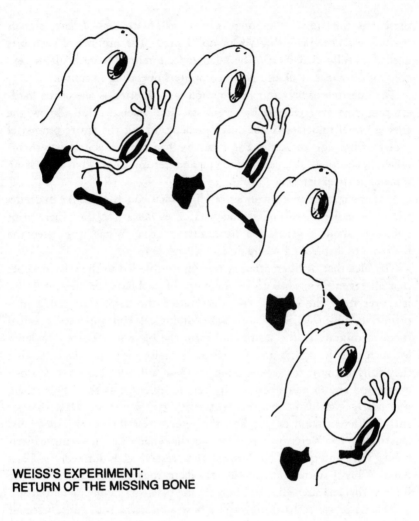

**WEISS'S EXPERIMENT:
RETURN OF THE MISSING BONE**

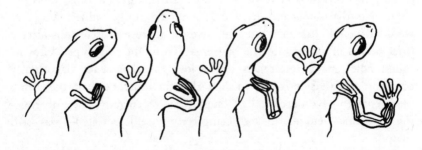

THE REVERSED-ARM EXPERIMENT

present is the digital computer, which deals with bits of data, signals that, in essence, say either yes or no, 1 or 0. The number of such bits needed to fully characterize the salamander forelimb is incalculable, exceeding the capacity of all known computers operating in unison.

The question of how this information is transferred is one of the hardest problems ever tackled by scientists, and when we fully know the answer, we'll understand not only regeneration but the entire process of growth from egg to adult. For now, we had best, as biologists themselves have done, skip this problem and return to it after addressing some slightly easier ones.

It seems reasonable that understanding what comes *out* of the blastema would be easier if we understood what goes *into* it, so the other major questions about regeneration have always been: What stimulates the blastema to form? And where do its cells come from?

The idea that dedifferentiation was impossible led to the related belief that all regeneration had to be the work of neoblasts, or "reserve cells" left over from the embryo and warehoused throughout the body in a primitive, unspecialized state. Some biological bell supposedly called them to migrate to the stump and form the blastema. There's evidence for such cells in hydras and flatworms, although it's now doubtful that they fully account for regeneration in these animals. However, no one ever found any in a salamander. In fact, as long ago as the 1930s, there was nearly conclusive evidence that they did *not* exist. Nevertheless, anti-dedifferentiation dogma and the reserve cell theory were defended fanatically, by Weiss in particular, so that many unconvincing experiments were interpreted to "prove" that reserve cells formed the blastema. When I started out, it was very dangerous for one's career even to suggest that mature cells might create the blastema by dedifferentiating.

Because it was so hard to imagine how a blastema could arise without dedifferentiation, the idea later developed that perhaps cells could *partially* dedifferentiate. In other words, perhaps muscle cells could become cells that *looked* primitive and completely unspecialized, but that would then take up their previous lives as mature muscle cells after a brief period of amnesia in the blastema. To fit the square peg into the round hole, many researchers did a lot of useless work, laboriously counting cell divisions to try to show that the muscle cells in the stump made enough new muscle cells to supply the regenerate. The embarrassing blastema—enigmatic and completely undifferentiated—was still there.

We now know (see Chapter 6) that at least some types of cells can revert completely to the primitive state and that such despecialization is

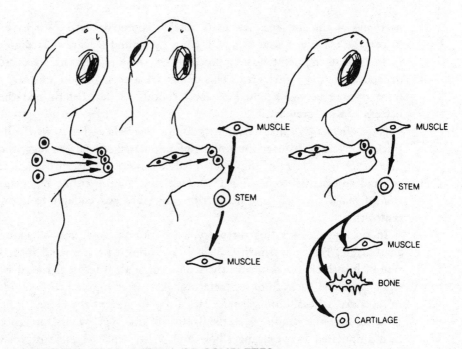

DEDIFFERENTIATION: PARTIAL OR COMPLETE?

the major, probably the only, way a blastema forms in complex animals
like a salamander.

Nerve Connections

The other major question about the blastema's origin is: What triggers
it? The best candidates for a "carrier" of the stimulus are the nerves. In
complex multicellular animals, there's no regeneration without nerve
tissue. Back in 1823, the English amateur Tweedy John Todd found
that if the nerves into a salamander's leg were cut when the amputation
was made, the limb wouldn't regrow. In fact, the stump itself shriveled
up and disappeared. However, Todd got normal regeneration when he
gave the nerves time to reconnect before severing the leg. Science wasn't
ready to make anything of his observation, but many experiments since
have confirmed it. Over a century later, Italian biologist Piera Locatelli
showed that an *extra* leg would grow if a nerve was rerouted so that it
ended near an intact leg. She cut the large sciatic nerve partway down
the salamander's hind leg, leaving it attached to the spinal column and
carefully threading it up under the skin so that its end touched the skin

near one of the forelegs. An extra foreleg sprouted there. When she placed the nerve end near a hind leg, an extra hind leg grew. It didn't matter where the nerve was supposed to be; the kind of extra structure depended on the target area. This indicated that some sort of energy from the nerves was adapted by local conditions that determined the pattern of what grew back.

Soon afterward, other researchers found that when they sewed full-thickness skin grafts over the stumps of amputated salamander legs, the dermis, or inner layer of the skin, acted as a barrier between the apical cap and an essential something in the leg, thereby preventing regeneration. Even a tiny gap in the barrier, however, was enough to allow regrowth.

In the early 1940s this discovery led S. Meryl Rose, then a young anatomy instructor at Smith College, to surmise that the rapid formation of full-thickness skin over the stumps of adult frogs' legs might be what prevented them from regenerating. Rose tried dipping the wounds in saturated salt solution several times a day to prevent the dermis from growing over the stump. It worked! Most of the frogs, whose forelimbs he'd amputated between the elbow and wrist, replaced some of what they'd lost. Several regrew well-formed wrist joints, and a few even began to produce one new finger. Even though the replacements were incomplete, this was a tremendously important breakthrough, the first time any regeneration had been artificially induced in an animal normally lacking the ability. However, the dermis *did* grow over the stump, so the experiment worked by some means Rose hadn't expected.

Later, other investigators showed that in normal regeneration the apical cap, minus the dermis, was important because regrowing nerve fibers made unique connections with the epidermal cells in the first stage of the process, before the blastema appeared. These connections are collectively called the neuroepidermal junction (NEJ). In a series of detailed experiments, Charles Thornton of Michigan State University cut the nerves to salamander legs at various times before amputating the legs, then followed the progress of the regrowing nerves. Regeneration began only after the nerves had reached the epidermis, and it could be prevented by any barrier separating the two, or started by any breach in the barrier. By 1954 Thornton had proved that the neuroepidermal junction was the one pivotal step that must occur before a blastema could form and regeneration begin.

Shortly thereafter, Elizabeth D. Hay, an anatomist then working at Cornell University Medical College in New York, studied the neuroepidermal junction with an electron microscope. She found that as

THE EPIDERMIS ALONE MUST GROW OVER THE END

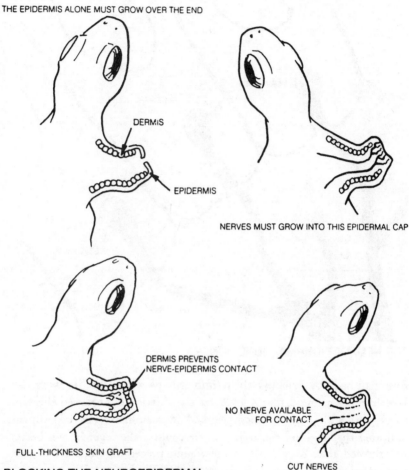

DERMIS

EPIDERMIS

NERVES MUST GROW INTO THIS EPIDERMAL CAP

DERMIS PREVENTS
NERVE-EPIDERMIS CONTACT

NO NERVE AVAILABLE
FOR CONTACT

FULL-THICKNESS SKIN GRAFT

CUT NERVES

**BLOCKING THE NEUROEPIDERMAL
JUNCTION PREVENTS REGENERATION**

each nerve fiber bundle reached the end of the stump, it broke up and each fiber went its separate way, snaking into the epidermis, which might be five to twenty cells thick. Each nerve fiber formed a tiny bulb at its tip, which was placed against an epidermal cell's membrane, nestling into a little pocket there. The arrangement was much like a synapse, although the microscopic structure wasn't as highly developed as in such long-term connections.

The junction was only a bridge, however. The important question was: What traffic crossed it?

In 1946, Lev Vladimirovich Polezhaev, a young Russian biologist then working in London, concluded a long series of experiments in which he induced partial regeneration in adult frogs, the same success

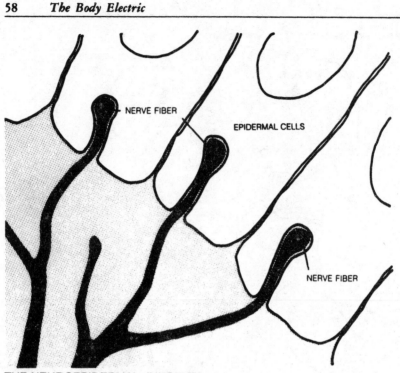

THE NEUROEPIDERMAL JUNCTION

Rose had had, by pricking their limb stumps with a needle every day. Polezhaev then found that a wide variety of irritants produced the same effect, although none of them worked in mammals. His experiments indicated that making the injury worse could make regeneration better, and showed that Rose's salt-in-the-wound procedure worked by irritation rather than by preventing dermis growth.

Next, the part that nerve tissue played was clarified considerably by Marcus Singer in a brilliant series of experiments at Harvard Medical School from the mid-1940s to the mid-1950s. Singer first confirmed Todd's long-forgotten work by cutting the nerves in salamander legs at various stages of regrowth, proving that the nerves were needed only in the first week, until the blastema was fully formed and the information transferred. After that, regeneration proceeded even if the nerves were cut.

Recent research had found that a salamander could replace its leg if all the motor nerves were cut, but not without the sensory nerves. Many assumed then that the growth factor was related only to sensory nerves, but Singer was uneasy over this conclusion: "The problem stated in advance that one *or* another nerve component is all important for regeneration." (Italics added.) Several facts didn't fit, however. Not only did the blastema fail to form when all nerves were cut, it didn't begin to form even if a substantial number, but still a minority, remained. Also, a salamander's leg would regrow with only motor nerves if extra motor nerves from the belly were redirected into the stump. In addition, zoologists had found that the sensory nerve contained more fibers than the motor nerve.

Singer counted for himself. In the thigh or upper arm, sensory fibers outnumbered motor by four to one. The ratio was even larger at the periphery. Then he cut them in various combinations in a long series of experiments. Regeneration worked as long as the leg had about one fourth to one third of its normal nerve supply, no matter in what combination. There seemed to be a threshold number of neurons (nerve cells) needed for regrowth.

But it wasn't that simple. The limbs of *Xenopus,* a South American frog unique in its ability to regenerate during adult life, had nerve fibers numbering well *under* the threshold. So Singer started measuring neuron *size,* and found that *Xenopus* had much bigger nerves than nonregenerating frogs. Another series of experiments verified the link: A critical mass—about 30 percent—of the normal nerve tissue must be intact for regeneration to ensue.

This finding made it pretty certain that whatever it was the nerves delivered didn't come from their known function of transmitting information by nerve impulses. If nerve impulses had been involved, regeneration should have faded away gradually with greater and greater flaws as the nerves were cut, instead of stopping abruptly when the minimum amount no longer remained.

Singer's discovery also provided a basic explanation for the decline of regeneration with increasing evolutionary complexity. The ratio between body mass and total nerve tissue is about the same in most animals, but more and more nerve became concentrated in the brain (a process called encephalization) as animals became more complex. This diminished the amount of nerve fiber available for stimulating regeneration in peripheral parts, often below the critical level.

In the early 1950s, Singer applied what he'd learned to the nonregenerating adult bullfrog. Using Locatelli's method, he dissected the sciatic nerve out of the hind leg, leaving it attached to the spinal cord, and directed it under the skin to the foreleg amputation stump. In two or three weeks, blastemas had formed, and the cut legs were restored to about the same degree as in Rose's and Polezhaev's experiments.

By 1954 Singer was ready to look for a growth-inducing chemical that was presumed to be coming from the nerves. The most promising

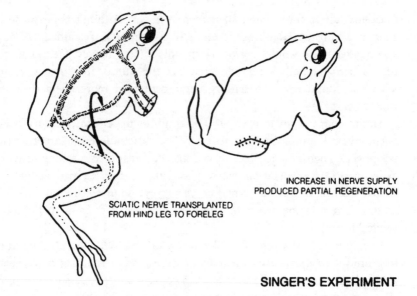

SCIATIC NERVE TRANSPLANTED
FROM HIND LEG TO FORELEG

INCREASE IN NERVE SUPPLY
PRODUCED PARTIAL REGENERATION

SINGER'S EXPERIMENT

possibility was the neurotransmitter acetylcholine, one of several compounds known to relay nerve impulses across synapses. The nerves se-
creted acetylcholine more abundantly than normal during blastema
formation—just when nerve supply was crucial—and its production fell
back to normal when regrowth was well under way. Singer had studied
previous failures with acetylcholine, in which experimenters had rubbed
it on the stump or injected it into the blastema. He thought these methods were too artificial, so he invented a microinfusion apparatus to release tiny amounts of acetylcholine continually, just as the nerves did. It
used a clock motor to drip the hormone slowly through a needle into the
shoulder of an anesthetized animal in which the leg nerves had been
removed. He had trouble keeping the drugged salamanders alive, so
maybe the anesthetic affected the outcome, but even the ones that survived didn't regenerate at all. The growth factor was almost certainly
not acetylcholine.

Vital Electricity

These, then, were the shoulders on which I stood in 1958 as I began to
look for the pattern-control and blastema-stimulating factors in regeneration. At that time we knew of two things that could yield some
regrowth in nonregenerators: extra nerve and extra injury. How were
they related? Luck gave me a clue.

I began my work just after the first few Sputniks, during the "missile gap" flap. Alarmed by the unforeseen triumphs of Russian technology, which we'd considered primitive, the government hastily began translating every Soviet scientific journal and distributing copies free to federally funded research centers. Suddenly, the medical library at the VA Medical Center in Syracuse, where I worked, began receiving each month a crate of Russian journals on clinical medicine and biology. Since no one else was much interested, this bonanza was all for me.

I soon made two discoveries: The Russians were willing to follow hunches; their researchers got government money to try the most outlandish experiments, ones that our science just *knew* couldn't work. Furthermore, Soviet journals published them—even if they *did* work. I particularly enjoyed *Biofizika,* the Soviet journal of biophysics, and it was there I encountered a paper on the "Nature of the Variation of the Bioelectric Potentials in the Regeneration Process of Plants," by A. M. Sinyukhin of Lomonosov State University in Moscow.

Sinyukhin began by cutting one branch from each of a series of tomato plants. Then he took electrical measurements around the wound as each plant healed and sent out a new shoot near the cut. He found a negative current—a stream of electrons—flowing from the wound for the first few days. A similar "current of injury" is emitted from all wounds in animals. During the second week, after a callus had formed over the wound and the new branch had begun to form, the current became stronger and reversed its polarity to positive. The important point wasn't the polarity—the position of the measuring electrode with respect to a reference electrode often determines whether a current registers as positive or negative. Rather, Sinyukhin's work was significant because he found a *change* in the current that seemed related to reparative growth. Sinyukhin found a direct correlation between these orderly electrical events and biochemical changes: As the positive current increased, cells in the area more than doubled their metabolic rate, also becoming more acidic and producing more vitamin C than before.

Sinyukhin then applied extra current, using small batteries, to a group of newly lopped plants, augmenting the regeneration current. These battery-assisted plants restored their branches up to three times faster than the control plants. The currents were very small—only 2 to 3 microamperes for five days. (An ampere is a standard unit of electric current, and a microampere is one millionth of an ampere.) Larger amounts of electricity killed the cells and had no growth-enhancing effect. Moreover, the polarity had to match that normally found in the plant. When Sinyukhin used current of the opposite polarity, nullifying the plant's own current, restitution was delayed by two or three weeks.

To American biology, however, this was all nonsense. To understand why, we must backtrack for a while.

Luigi Galvani, an anatomy professor in the medical school at the University of Bologna who'd been studying electricity for twenty years, first discovered the current of injury in 1794, but unfortunately he didn't know it.

At that time, biology's main concern was the debate between vitalism and mechanism. Vitalism, though not always called by that name, had been the predominant concept of life since prehistoric times throughout the world, and it formed the basis for almost all religions. It was closely related to Socrates' and Plato's idea of supernatural "forms" or "ideals" from which all tangible objects and creatures derived their individual characteristics. Hippocrates adapted this idea by postulating an anima as the essence of life. The Platonic concept evolved into the medieval philosophy of realism, whose basic tenet was that abstract universal principles were more real than sensory phenomena. Mechanism grew out of Aristotle's less speculative rationalism, which held that universal principles were *not* real, being merely the names given to humanity's attempts at making sense of the reality apprehended through the senses. Mechanism had become the foundation of science through the writings of Descartes in the previous century, although even he believed in an "animating force" to give the machine life at the outset. By Galvani's time, mechanism's influence was steadily growing.

Galvani was a dedicated physician, and medicine, tracing its lineage back to tribal shamans, has always been a blend of intuition and empirical observation based on a vitalistic concept of the sanctity of life. The vitalists had long tried—unsuccessfully—to link the strange, incorporeal phenomenon of electricity with the *élan vital*. This was Galvani's main preoccupation.

One day he noticed that some frogs' legs he'd hung in a row on his balustrade, pending his dinner, twitched whenever the breeze blew them against the ironwork. At about the same time his wife Lucia noticed in his laboratory that the muscles of a frog's leg contracted when an assistant happened to be touching the main nerve with a steel scalpel at the same instant that a spark leaped from one of the electrical machines being operated across the room. (The only type of electricity then known was the static type, in the form of sparks from various friction devices.) Today we know that an expanding and collapsing electric field generated by the spark induced a momentary current in the scalpel, which stimulated the muscle, but Galvani believed that the metal railing and scalpel had drawn forth electricity hidden in the nerves.

Galvani experimented for years with nerves from frogs' legs, connected in various circuits with several kinds of metals. He grew convinced that the vital spirit was electricity flowing through the nerves and announced this to the Bologna Academy of Science in 1791.

Within two years, Alesandro Volta, a physicist at the University of Padua, had proven that Galvani had in fact discovered a new kind of electricity, a steady current rather than sparks. He'd generated a bimetallic direct current, a flow of electrons between two metals, such as the copper hooks and iron railing of the famous balcony observation, connected by a conducting medium—in other words, a battery. The frogs' legs, being more or less bags of weak salt solution, were the electrolyte, or conducting medium. They were otherwise incidental, Volta explained, and there was no such thing as Galvani's "animal electricity."

Galvani, a shy and thoroughly noncombative soul, was crushed. His only response was an anonymous paper in 1794 describing several experiments in which frogs' legs could be made to twitch with no metal in the circuit. In one procedure, the experimenter touched one leg nerve with the frog's dissected-out, naked spinal cord, while holding the other leg to complete the circuit. Here the current was true animal electricity, coming from the amputation wound at the base of the leg.

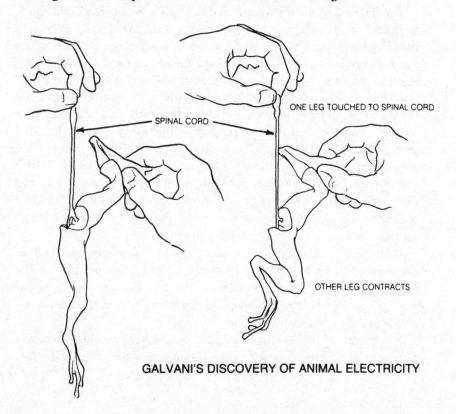

SPINAL CORD

ONE LEG TOUCHED TO SPINAL CORD

OTHER LEG CONTRACTS

GALVANI'S DISCOVERY OF ANIMAL ELECTRICITY

In the long run, Galvani unwittingly helped the cause of the mechanists by giving them something to attack. As long as the *élan vital* was ephemeral, all you could say was that you couldn't find it. Once Galvani said it was electricity, a detectable, measurable entity, there was a target for experimentation. Actually, Baron Alexander von Humboldt, the explorer-naturalist who founded geology, proved in 1797 that Volta and Galvani were both partly right. Bimetallic currents were real, but so was spontaneous electricity from injured flesh. However, the mechanists had the upper hand; Galvani's anonymous report and Humboldt's confirmation were overlooked. Galvani himself died penniless and brokenhearted in 1798, soon after his home and property were confiscated by the invading French, while Volta grew famous developing his storage batteries under the auspices of Napoleon.

Then in the 1830s a professor of physics at Pisa, Carlo Matteucci, using the newly invented galvanometer, which could measure fairly small direct currents, came up with other evidence for animal electricity. In a meticulous series of experiments lasting thirty-five years, he conclusively proved that the current of injury was real. However, he didn't find it in the nervous system, only emanating from the wound surface, so it couldn't be firmly related to the vital force.

The tale took another turn in the 1840s when Emil Du Bois-Reymond, a physiology student in Berlin, read Matteucci's work. Du Bois-Reymond went on to show that when a nerve was stimulated, an impulse traveled along it. He measured the impulse electrically and announced his conclusion that it was a mass of "electromotive particles," like a current in a wire. Immediately he squared his shoulders, expecting the mantle of glory to descend: "If I do not greatly deceive myself," he wrote, "I have succeeded in realizing in full actuality . . . the hundred years' dream of physicists and physiologists, to wit, the identity of the nervous principle with electricity." But he *had* deceived himself. Soon it was learned that the impulse traveled too slowly to be a current, and that nerves didn't have the proper insulation or resistance to conduct one, anyway. Any true current the size of the small measured impulse wouldn't have made it through even a short nerve.

Julius Bernstein, a brilliant student of Du Bois-Reymond, resolved the impasse in 1868 with his hypothesis of the "action potential." The impulse wasn't a current, Bernstein said. It was a disturbance in the ionic properties of the membrane, and it was this perturbation that traveled along the nerve fiber, or axon.

The Bernstein hypothesis stated that the membrane could selectively filter ions of different charges to the inside or outside of the cell. (Ions are charged particles into which a salt breaks up when dissolved in water; all salts dissociate in water into positive and negative ions, such as the positive sodium and negative chloride ions of table salt.) Bernstein postulated that the membrane could sort most of the negatives outside and most of the positives inside the fiber. The membrane was polarized (with like charges grouped on one side), having a transmembrane potential, because the negative charges, all on one side, could potentially flow in a current across the membrane to achieve a balance on both sides. This was what happened in a short segment of the membrane whenever a nerve was stimulated. Part of the membrane became depolarized, reversing the transmembrane potential. The nerve impulse was actually a disturbance in the potential traveling along the membrane. As the area of disturbance moved along, the membrane quickly restored its normal resting potential. Thus the nerve impulse wasn't an electrical current, even though it could be measured electrically.

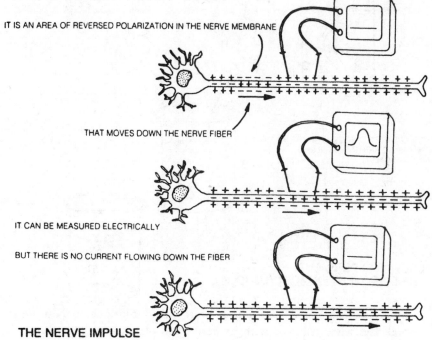

IT IS AN AREA OF REVERSED POLARIZATION IN THE NERVE MEMBRANE

THAT MOVES DOWN THE NERVE FIBER

IT CAN BE MEASURED ELECTRICALLY

BUT THERE IS NO CURRENT FLOWING DOWN THE FIBER

THE NERVE IMPULSE

Bernstein's hypothesis has been confirmed in all important respects, although it remains a hypothesis because no one has yet found what gives the membrane the energy to pump all those ions back and forth. Soon it was broadened, however, to include an explanation of the current of injury. Reasoning that all cells had transmembrane potentials, Bernstein maintained that, after injury, the damaged cell membranes simply leaked their ions out into the environment. Thus the current of injury was no longer a sign that electricity was central to life, but only an uninteresting side effect of cell damage.

The vitalists, with their hopes pinned on electricity, kept getting pushed into tighter and tighter corners as electricity was removed from one part of the body after another. Their last stand occurred with the discovery of neurotransmitters. They'd maintained that only an electrical current could jump across the synapse, the gap between communicating nerves. In 1920 that idea was disproven with a lovely experiment by Otto Loewi, a research professor at the NYU School of Medicine, later my alma mater. When I took physiology in my first year there, we had to duplicate his experiment.

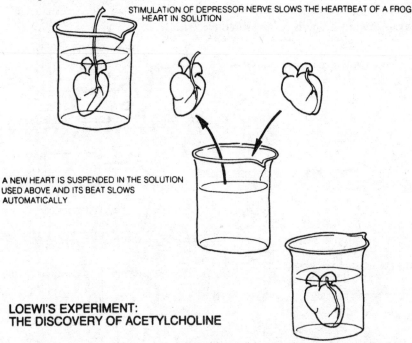

STIMULATION OF DEPRESSOR NERVE SLOWS THE HEARTBEAT OF A FROG HEART IN SOLUTION

A NEW HEART IS SUSPENDED IN THE SOLUTION USED ABOVE AND ITS BEAT SLOWS AUTOMATICALLY

LOEWI'S EXPERIMENT:
THE DISCOVERY OF ACETYLCHOLINE

Biologists had found that a frog heart would continue to beat for several days when removed with its nerves and placed in an appropriate solution. Stimulating one of the nerves would slow it down. Like Loewi, we took one such heart, with nerve attached, and stimulated the nerve, slowing the beat. We then collected the solution bathing that heart and placed another heart in it. Its beat slowed even though its depressor nerve hadn't been stimulated. Obviously the nerve slowed the heartbeat by producing a chemical, which crossed the gap between the nerve ending and the muscle fiber. This chemical was later identified as acetyl-choline, and Loewi was awarded the Nobel Prize in 1936 for this discovery. His work resulted in the collapse of the last vestige of electrical vitalism. Thereafter, every function of the nervous system had to be explained on the basis of the Bernstein hypothesis and chemical transmission across the synapse.

It was with great trepidation, therefore, that I put any credence in Sinyukhin's report that the strength of the injury current affected regeneration in his plants. Yet his report was detailed and carefully written. Something about his work gave me a gut feeling that it was valid. Maybe it was because the tomato plants he used were "Best of All" American Beauties. At this point I wasn't aware of Matteucci's forgotten work, but something clicked in my mind now as I studied Rose's and Polezhaev's experiments. In both, definitely in Polezhaev's and probably in Rose's, regeneration had been stimulated by an increase in the injury.

Then another Russian supplied a timely lead. In a government translation I found a 1958 paper by A. V. Zhirmunskii of the Institute of Cytology in Leningrad, who studied the current of injury in the hind leg muscle of the bullfrog. This muscle is nice and long, easy to work with, and contains branches from several different nerves. He made a standard injury in each muscle, measured the current of injury, then cut the nerves branch by branch, noting the effect on the current. It decreased with each succeeding nerve cut. The current of injury was proportional to the amount of nerve.

Then I went to the library and delved back into the history of neurophysiology and found Matteucci's superb series of observations. Not only had he proven that the current of injury was real, he'd shown that it varied in proportion to the severity of the wound.

Now I had enough pieces to start on the puzzle. I summarized the observations in a little matrix:

Extent of injury is proportional to regeneration
Amount of nerve is proportional to regeneration
Extent of injury is proportional to current of injury
Amount of nerve is proportional to current of injury
Ergo: current of injury is proportional to regeneration

I was pretty sure now that, contemporary "knowledge" to the contrary, the current of injury was no side effect and was the first place to look for clues to the growth-control and dedifferentiation-stimulating factors. I planned my first experiment.

Three

The Sign of the Miracle

Real science is creative, as much so as painting, sculpture, or writing. Beauty, variously defined, is the criterion for art, and likewise a good theory has the elegance, proportion, and simplicity that we find beautiful. Just as the skilled artist omits the extraneous and directs our attention to a unifying concept, so the scientist strives to find a relatively simple order underlying the apparent chaos of perception. Perhaps because it was mine, my theory that the current of injury stimulated regeneration seemed both simple and beautiful. It's impossible to convey the sense of excitement I felt when all of the facts fell together and the idea came. I'd created something new that explained the previously inexplicable. I couldn't wait to see if I was right.

In all the time that the Bernstein hypothesis had been used to explain away the current of injury, no one had ever thought to measure the current over a period of days to see how long it lasted. If it was only ions leaking from damaged cells, it should disappear in a day or two, when these cells had finished dying or repairing themselves. This simple measurement, with a comparison of the currents in regenerating versus nonregenerating limbs, was what I planned to do. I would uniformly amputate the forelegs of frogs and salamanders. Then, as the frogs' stumps healed over and the salamanders' legs regrew, I would measure the currents of injury each day.

The experiment itself was as simple as could be. The tricky part was getting permission to do it.

When you want to do a research project, there are certain channels you must go through to get the money. You write a project proposal, spelling out what hypothesis you want to test, why you think it should be done, and how you plan to go about it. The proposal goes to a committee supposedly composed of your peers, people who have demonstrated competence in related research. If they approve your project and the money is available, you generally get part of what you asked for, enough to get started.

The Veterans Administration had been dispensing research money for several years as a sort of bribe to attract doctors despite the low pay in government service. The money from Washington was doled out by the most influential doctors on the staff, not necessarily the best researchers, but I still felt I had a good chance because the VA was having an especially hard time recruiting orthopedists. Moreover, my hypothesis was based on the work of Rose, Polezhaev, Singer, Sinyukhin, and Zhirmunskii with inescapable logic. And since frogs and salamanders were anatomically similar, any difference in their currents of injury should reflect the disparity in their powers of regeneration. My chances of being thrown off by extraneous factors were thus minimal.

I remember thinking, as I wrote the proposal, how my life had come full circle. As a college freshman in 1941, I'd conducted a crude experiment on salamanders, showing that thyroid stimulation by iodine didn't speed up regeneration. Here I was nearly twenty years later, beneficiary of the intervening research, hoping to add to our knowledge of the same phenomenon and perhaps even discover something that would help human patients. I worried that my roundabout course might weigh against me, since one of the criteria for grants was whether the investigator had been trained for that particular field. This proposal would have been expected from a physiologist, not an orthopedist. Nevertheless, I was asking for a relatively minuscule amount of money. I needed only a thousand dollars to put together the equipment, so I didn't anticipate much trouble.

The Tribunal

"Dr. Becker, could you please come to a special research committee meeting in one hour?" The committee's secretary was calling. I'd known something was up; two months had passed since I'd filed my proposal, and all my queries as to its fate had gone unanswered.

"I'll be there."

"It's not here in the research office. It's downstairs in the hospital director's office." Now that was really strange. The director almost never paid any attention to the research program. Besides, his office was big enough to hold a barbecue in.

It was a barbecue, ali right, and I was the one being grilled. The director's conference room had been rearranged. In place of the long, polished table there was a semicircle of about a dozen chairs, each occupied by one of the luminaries from the hospital and medical school. I recognized the chairmen of the departments of biochemistry and physiology along with the hospital director and chief of research. Only the dean was missing. In the center was a single chair—for me.

The spokesman came right to the point: "We have a very grave basic concern over your proposal. This notion that electricity has anything to do with living things was totally discredited some time ago. It has absolutely no validity, and the new scientific evidence you're citing is worthless. The whole idea was based on its appeal to quacks and the gullible public. I will not stand idly by and see this medical school associated with such a charlatanistic, unscientific project." Murmurs of assent spread around the group.

I had the momentary thrill of imagining myself as Galileo or Giordano Bruno; I thought of walking to the window to see if the stake and fagots were set up on the lawn. Instead I delivered a terse speech to the effect that I still thought my hypothesis was stoutly supported by some very good research and that I was sorry if it flew in the face of dogma. I ended by saying that I didn't intend to withdraw the proposal, so they would have to act upon it.

When I got home, my fury was gone. I was ready to call the director, withdraw my proposal, apologize for my errors, stay out of research, quit the VA, and go into private practice, where I could make a lot more money. Luckily, my wife Lil knows me better than I sometimes know myself. She told me, "You'd be miserable in private practice. This is exactly what you want to do, so just wait and see what happens."

Two days later I got word that the committee had delegated the decision to Professor Chester Yntema, an anatomist who long ago had studied the regrowth of ears in the salamander. Since he was the only one in Syracuse who'd ever done any regeneration research, I've always wondered why he wasn't part of the first evaluation. I went to see him with a sense of foreboding, for his latest research seemed to refute Singer's nerve work, on which I'd based my proposal.

Using a standard technique, Yntema had operated on very young salamander embryos, cutting out all of the tissues that would have given rise to the nervous system. He then grafted each of these denervated embryos onto the back of a whole one. The intact embryos furnished the grafts with blood and nourishment, and the procedure resulted in a little "parabiotic" twin, normal except for having no nerves, stuck on the back of each host animal. Yntema then cut off one leg from each of these twins, and some of them regenerated. Since microscopic examination revealed no nerves entering the graft from its host, Yntema's experiment called Singer's conclusions into question.

Dr. Yntema turned out to be one of the nicest gentlemen I've ever met, but as I entered his office his Dickensian appearance of eminence— he was tall, thin, elderly, with craggy features, and wore an immaculately starched, long white lab coat—made me feel like a freshman being called before the dean. But he put me at ease immediately.

"I've read your proposal and think it's most intriguing," he said with genuine interest.

"Do you really?" I asked. "I've been afraid you would reject it out of hand because my ideas depend on Singer's work."

"Marc Singer is a good, careful worker," Yntema replied. "I don't doubt his observations. What I've described is an exception to his findings under special circumstances."

After a long, pleasant conversation about regrowth, nerves, and research itself, he gave me his approval with a word of caution: "Don't get your hopes up about what you want to do. I don't believe for one minute that it'll work, but I think you should do it anyway. We need to encourage young researchers. Besides, it'll be fun, and maybe you'll learn something new, after all. Let me know what happens, and if you need any help, I'll be here. I'll call the people at the VA right away, so get to work. Good luck."

This was the start of a long friendship. I'm deeply indebted to Chester Yntema for his encouragement. Had he not believed that research should be fun, that you should do what you want rather than what's fashionable, my first experiment would have been impossible, and this book would never have been written.

The Reversals

First I found a good supplier of salamanders and frogs, a Tennessee game warden who ran this business in his spare time. Sometimes the shipment would contain a surprise, a small snake. I never found out whether he included them deliberately or by error. At any rate, his animals weren't the inferior aquarium-bred stock but robust specimens collected from their natural habitats.

Next I worked out some technical problems. The most important of these was the question of where to place the electrodes. To form the circuit, two electrodes had to touch the animal. One was the "hot" or measuring electrode, which determined the polarity, positive or negative, with regard to a stationary reference electrode. A negative polarity meant there were more electrons where the measuring electrode was placed, while a positive polarity meant there were more at the reference site. A steady preponderance of negative charge at a particular location could mean there was a current flowing toward that spot, continually replenishing the accumulation of electrons. The placement of the reference electrode, therefore, was critical, lest I get the voltage right but the polarity, and hence the direction of the current, wrong. Some logical position had to be chosen and used every time. Since I postulated that the nerves were somehow related to the current, the cell bodies that sent their nerve fibers into the limb seemed like a good reference point. These cell bodies were in a section of the spinal cord called the brachial enlargement, located just headward from where the arm joined the body. In both frogs and salamanders, therefore, I put the measuring electrode directly on the cut surface of the amputation stump and the reference electrode on the skin over the brachial enlargement.

After setting up the equipment, I did some preliminary measurements on the intact animals. They all had areas of positive charge at the brachial enlargement and a negative charge of about 8 to 10 millivolts at each extremity, suggesting a flow of electrons from the head and trunk out into the limbs and, in the salamanders, the tail.

I began the actual experiment by amputating the right forelimbs, between elbow and wrist, from fourteen salamanders and fourteen grass frogs, all under anesthesia. I took no special precautions against bleeding, since blood clots formed very rapidly. The wounds had to be left open, not only because closing the skin over the salamanders' amputation sites would have stopped regeneration, but also because I was investigating a natural process. In the wild, both frogs and salamanders get injuries much like the one I was producing—both are favorite foods of the freshwater bass—and heal them without a surgeon.

Once the anesthetic wore off and the blood clot formed, I took a voltage reading from each stump. I was surprised to find that the polarity at the stump reversed to positive right after the injury. By the next day it had climbed to over 20 millivolts, the same in both frogs and salamanders.

I made measurements daily, expecting to see the salamander voltages climb above those of the frogs as the blastemas formed. It didn't work that way. The force of the current flowing from the salamanders' amputation sites rapidly dropped, while that from the frogs' stumps stayed at the original level. By the third day the salamanders showed no current at all, and their blastemas hadn't even begun to appear.

The experiment seemed a failure. I almost quit right there, but something made me keep on measuring. I guess I thought it would be good practice.

Then, between the sixth and tenth days an exciting trend emerged. The salamander potentials changed their sign again, exceeding their normal voltage and reaching a peak of more than 30 millivolts negative just when the blastemas were emerging. The frogs were still plugging away with slowly declining positive voltages. As the salamander limbs regenerated and the frog stumps healed over with skin and scar tissue, both groups of limbs gradually returned (from opposite directions) to the original baseline of 10 millivolts negative.

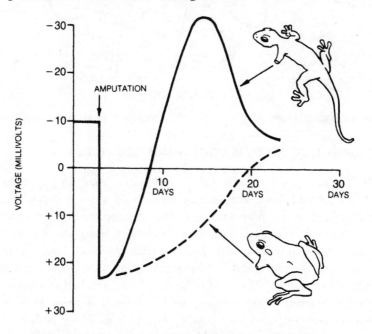

THE CURRENT OF INJURY: SALAMANDER VERSUS FROG

Here was confirmation better than my wildest dreams! Already, in my first experiment, I had the best payoff research can give—the excitement of seeing something no one else had ever seen before. I knew now that the current of injury wasn't due to dying cells, which were long gone by then. Moreover, the opposite polarities indicated a profound difference in the electrical properties of the two animals, which somehow would explain why only the salamander could regenerate. The negative potential seemed to bring forth the all-important blastema. It was a very significant observation, even though the facts had scrambled my neat hypothesis somewhat.

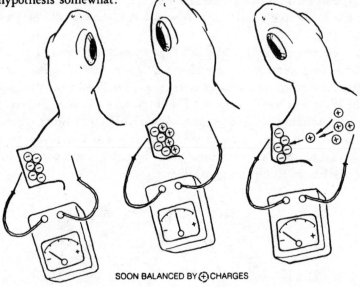

SOON BALANCED BY ⊕ CHARGES

SIMPLE ⊖ CHARGES FROM DAMAGED CELLS   THERE MUST BE A CONTINUOUS CURRENT

THE CURRENT OF INJURY IS MORE THAN A SIDE EFFECT

Dr. Yntema agreed and urged me to write up a report for publication, but first I jumped ahead with another idea. I took a new group of frogs, amputated one foreleg from each, and every day applied negative current to the stump from a small battery. I dreamed of being the first to get complete regrowth in a normally nonregenerating animal; I could almost see my name on the cover of *Scientific American*. The frogs were less interested in my glory. They had to hold still for up to half an hour with electrodes attached. They refused, so I anesthetized them each day, something they tolerated very poorly. Within a week my Nobel Prize had turned into a collection of dead frogs.

For some time I'd been scouring the dusty stacks of the medical library for previous work on bioelectricity, and now I found a paper written in 1909 by an American researcher named Owen E. Frazee. He reported that electrical currents passed through the aquarium water in which larval salamanders were living speeded up their regeneration. At that time, electrical equipment was so primitive that I couldn't rely on Frazee's results, but I decided to try it for myself. What Sinyukhin had done with tomato plants I hoped to do with salamanders.

To one group of salamanders I applied 2 microamperes of positive current from batteries connected directly to the stumps for five to ten minutes on each of the first five days after amputation. This was 0.000002 ampere, a tiny current by ordinary standards (most household circuits carry 15 or 20 amperes) but comparable to what seemed to be flowing in the limb. I intended to reinforce the normal positive peak in the current of injury. This treatment seemed to make the blastemas larger but slowed down the whole process somewhat. To another group I applied 3 microamperes of negative current on the fifth to ninth days, when the normal currents were hitting their negative peaks. This seemed to increase the rate of regrowth for a week but didn't change the time needed for a complete limb. Finally I tried Frazee's method with a constant current through the aquarium water. Again the results were equivocal at best. These failures taught me that, before I applied my findings to other animals, I would have to learn *how* the current of injury worked.

Meanwhile, I wrote up my results. Not knowing any better, I submitted my paper to the *Journal of Bone and Joint Surgery*, the most prestigious orthopedic journal in the world. It was a dumb thing to do. The experiment had no immediate practical application, while the journal accepted only clinical reports. Moreover, the publication was very political; normally you had to have an established reputation or come from one of the big orthopedic programs, like Harvard or Columbia, to get into it. Luckily, I didn't know that. Someone thought my paper was just what the doctor ordered. Not only was it accepted for publication, but I was invited to present it at the next combined meeting of the Orthopaedic Research Society and the American Academy of Orthopaedic Surgeons, at Miami Beach in January 1961. This invitation was a particular honor, for it meant someone considered my work so significant that practicing physicians, as well as researchers, should hear of it right then and there. Whoever that someone was, he or she has my undying gratitude.

My report was well received and soon was published, to the consternation of the local inquisitors and the delight of Chester Yntema. Since the journal was geared to clinicians, I worried that my experiment wouldn't reach the basic researchers with whom I really wanted to share it, but again I was wrong. The next year I got a phone call from Meryl

Rose himself. He was excited by the article and wanted to know what I'd done since.

Although Rose taught at Tulane Medical School in New Orleans, he spent every summer at the Woods Hole Marine Biological Laboratory on Cape Cod, so he and his wife drove to Syracuse from there. Despite his success, Rose had maintained the completely open mind that a great researcher must have, and he was fascinated by the observations on electric fields, nerves, anesthesia, and magnetism that I'll recount in the next chapter. Since then his interest has encouraged me enormously. My friendship with this fine man and scientist has been fruitful even beyond the expectations I had then, and, when my wife and I had the Roses to dinner, we found our pasts were linked by an odd coincidence. As they walked in the door, Lillian exclaimed, "Dr. Rose! Weren't you at Smith College in the 1940s?" It turned out that she'd been a friend of Rose's student lab assistant and had helped catch the frogs for the famous salt-in-the-wound experiment!

Part 2

The Stimulating Current

The basic texture of research consists of dreams into which
the threads of reasoning, measurement, and calculation
are woven.

—ALBERT SZENT-GYÖRGYI

Four

Life's Potentials

It's an axiom of science that the better an experiment is, the more new questions it raises after it has answered the one you asked. By that standard my first simple test had been pretty good. The new problems branched out like the fingers on those restored limbs: Where did the injury currents come from? Were they in fact related to the nervous system and, if so, how? It seemed unlikely that they sprang into action only after an amputation; they must have existed before. There must have been a preexisting substratum of direct current activity that responded to the injury. Did the voltages I measured really reflect such currents, and did they flow throughout the salamander's body? Did other organisms have them? What structures carried them? What were their electrical properties? What were they doing the rest of the time, before injury and after healing? Could they be used to provoke regeneration where it was normally absent?

I had ideas about how to look for some of the answers, but, to understand my approach, the reader unfamiliar with electrical terms will need a simplified explanation of several basic concepts that are essential to the rest of the story.

Everything electrical stems from the phenomenon of charge. No one knows exactly what this is, except to say that it's a fundamental property of matter that exists in two opposite forms, or polarities, which we arbitrarily call positive and negative. Protons, which are one of the two main types of particles in atomic nuclei, are positive; the other particles, the neutrons, are so named because they have no charge. Orbiting around the nucleus are electrons, in the same number as the protons inside the nucleus. Although an electron is 1,836 times less massive than a proton, the electron carries an equal but opposite (negative) charge. Because of their lightness and their position outside the nucleus, electrons are much more easily dislodged from atoms than are protons, so they're the main carriers of electric charge. For the lay person's purposes a negative charge can be thought of as a surplus of electrons, while a positive charge can be considered a scarcity of them. When electrons move away from an area, it becomes positively charged, and the area to which they move becomes negative.

A flow of electrons is called a current, and is measured in amperes, units named for an early-nineteenth-century French physicist, André Marie Ampère. A direct current is a more or less even flow, as opposed to the instantaneous discharge of static electricity as sparks or lightning, or the back-and-forth flow of alternating current which powers most of our appliances.

Besides the amount of charge being moved, a current has another characteristic important for our narrative—its electromotive force. This can be visualized as the "push" behind the current, and it's measured in volts (named for Alessandro Volta).

In high school most of us learned that a current flows only when a source of electrons (negatively charged material) is connected to a material having fewer free electrons (positively charged in relation to the source) by a conductor, through which the electrons can flow. This is what happens when you connect the negative terminal of a battery to its positive pole with a wire or a radio's innards: You've completed a circuit between negative and positive. If there's no conductor, and hence no circuit, there's only a hypothetical charge flow, or electric potential, between the two areas. The force of this latent current is also measured in volts by temporarily completing the circuit with a recording device, as I did in my experiment.

The potential can continue to build until a violent burst of current equalizes the charges; this is what happens when lightning strikes. Smaller potentials may remain stable, however. In this case they must be continuously fed by a direct current flowing from positive to negative, the opposite of the normal direction. In this part of a circuit, electrons actually flow from where they're scarce to where they're more abundant. As Volta found, such a flow is generated inside a battery by the electrical interaction of two metals.

An electric field forms around any electric charge. This means that any other charged object will be attracted (if the polarities are opposite) or repelled (if they're the same) for a certain distance around the first object. The field is the region of space in which an electrical charge can be detected, and it's measured in volts per unit of area.

Electric fields must be distinguished from magnetic fields. Like charge, magnetism is a dimly understood intrinsic property of matter that manifests itself in two polarities. Any flow of electrons sets up a combined electric and magnetic field around the current, which in turn affects other electrons nearby. Around a direct current the electromagnetic field is stable, whereas an alternating current's field collapses and reappears with its poles reversed every time the current changes direction. This reversal happens sixty times a second in our normal house currents. Just as a current produces a magnetic field, a magnetic field, when it moves in relation to a conductor, induces a current. Any varying magnetic field, like that around household appliances, generates a current in nearby conductors. The weak magnetic fields we'll be discussing are measured in gauss, units named after a nineteenth-century German pioneer in the study of magnetism, Karl Friedrich Gauss.

Both electric and magnetic fields are really just abstractions that scientists have made up to try to understand electricity's and magnetism's action at a distance, produced by no known intervening material or energy, a phenomenon that used to be considered impossible until it became undeniable. A field is represented by lines of force, another abstraction, to indicate its direction and shape. Both kinds of fields decline with distance, but their influence is technically infinite: Every time you use your toaster, the fields around it perturb charged particles in the farthest galaxies ever so slightly.

In addition, there's a whole universe full of electromagnetic energy, radiation that somehow seems to be both waves in an electromagnetic field and particles at the same time. It exists in a spectrum of wavelengths that includes cosmic rays, gamma rays, X rays, ultraviolet radiation, visible light, infrared radiation, microwaves, and radio waves. Together, electromagnetic fields and energies interact in many complex ways that have given rise to much of the natural world, not to mention the whole technology of electronics.

You'll need a casual acquaintance with all these terms for the story ahead, but don't worry if the concepts seem a bit murky. Physicists have been trying for generations to solve the fundamental mysteries of electromagnetism, and no one, not even Einstein, has yet succeeded.

Unpopular Science

None of these things had the slightest relevance to life, according to most biologists around 1960. A major evaluation of American medicine, financed by the Carnegie Foundation and published in 1910 by the respected educator Abraham Flexner, had denounced the clinical use of electric shocks and currents, which had been applied, often over-enthusiastically, to many diseases since the mid-1700s. Electrotherapy sometimes seemed to work, but no one knew why, and it had gotten a bad name from the many charlatans who'd exploited it. Its legitimate proponents had no scientific way to defend it, so the reforms in medical education that followed the Flexner report drove all mention of it from the classroom and clinic, just as the last remnants of belief in vital electricity were being purged from biology by the discovery of acetylcholine. This development dovetailed nicely with expanding knowledge of biochemistry and growing reliance on the drug industry's products. Penicillin later made medicine almost exclusively drug oriented.

Meanwhile, the work of Faraday, Edison, Marconi, and others literally electrified the world. As the uses of electricity multiplied, no one found any obvious effects on living creatures except for the shock and heating caused by large currents. To be sure, no one looked very hard, for fear of discouraging a growth industry, but the magic of electricity seemed to lie precisely in the way it worked its wonders unseen and unfelt by the folks clustered around the radio or playing cards under the light bulb. By the 1920s, no scientist intent on a respectable career dared suggest that life was in any sense electrical.

Nevertheless, some researchers kept coming up with observations that didn't fit the prevailing view. Although their work was mostly consigned to the fringes of the scientific community, by the late 1950s they'd accumulated quite a bit of evidence.

There were two groups of dissenters, but, because their work went unheeded, each was largely unaware of the other's existence. One line of inquiry began just after the turn of the century when it was learned that hydras were electrically polarized. The head was found to be positive, the tail negative. I've already mentioned Frazee's 1909 report of salamander regeneration enhanced by electrical currents. Then, with a classic series of experiments in the early 1920s, Elmer J. Lund of the University of Texas found that the polarity of regeneration in species related to the hydra could be controlled, even reversed, by small direct currents passed through the animal's body. A current strong enough to override the creature's normal polarity could cause a head to form where a tail should have reappeared, and vice versa. Others confirmed this discovery, and Lund went on to study eggs and embryos. He claimed to have influenced the development of frog eggs not only with currents but also with magnetic fields, a conclusion that was *really* risqué for that time.

Stimulated by Lund's papers, Harold Saxton Burr of Yale began putting electrodes to all kinds of creatures. Burr was lucky enough to have a forum for his work. He edited the *Yale Journal of Biology and Medicine,* where most of his reports appeared; few other journals would touch them. Burr and his co-workers found electric fields around, and electric potentials on the surfaces of, organisms as diverse as worms, hydras, salamanders, humans, other mammals, and even slime molds. They measured changes in these potentials and correlated them to growth, regeneration, tumor formation, drug effects, hypnosis, and sleep. Burr claimed to have measured field changes resulting from ovulation, but others got contradictory results. He hooked up his voltmeters to trees for years at a time and found that their fields varied in response not only to light and moisture, but to storms, sunspots, and the phases of the moon as well.

Burr and Lund were handicapped by their instruments as well as the research climate. Most of their work was done before World War II and, even though Burr spent years designing the most sensitive devices possible using vacuum tubes, the meters were still too "noisy" to reliably measure the tiny currents found in living things. The two scientists could refine their observations only enough to find a simple dipolar distribution of potentials, the head of most animals being negative and the tail positive.

Burr and Lund advanced similar theories of an electrodynamic field, called by Burr the field of life or L-field, which held the shape of an organism just as a mold determines the shape of a gelatin dessert. "When we meet a friend we have not seen for six months there is not one molecule in his face which was there when we last saw him," Burr wrote. "But, thanks to his controlling L-field, the new molecules have fallen into the old, familiar pattern and we can recognize his face."

Burr believed that faults in the field could reveal latent illness just as dents in a mold show up in the jelly. He claimed to be able to predict all sorts of things about a person's emotional and physical health, both present and future, merely by checking the voltage between head and hand. His later writings were marred by a sort of bioelectric determinism and a tendency to confuse "law and order" in nature with that odious euphemism as preached by Presidents. As a result, he began to suggest his simple readings as a foolproof way to evaluate job applicants, soldiers, mental patients, and suspected criminals or dissidents.

The fields Burr and Lund found were actually far too simple to account for a salamander's limb or a human face. Biological knowledge at that time gave them no theoretical framework to explain where their fields came from. They conceived of currents flowing within cells but had no proof. They had no inkling that currents might flow in specific tissues or in the fluids outside cells. They suggested that all these little intracellular currents somehow added up to the whole field. Burr wrote that "electrical energy is a fundamental attribute of protoplasm and is an expression or measure of the presence of an electrodynamic field in the organism." Unfortunately, an analysis of this sentence yields nonsense, and Burr's work was dismissed as foggy vitalism. Lund suffered the same fate. No one bothered to see if the *measurements* they'd made were valid. After all, you can disagree with a theory, but you should respect the data enough to check them. If you can't duplicate them, you're entitled to rest easy with your own concepts, but if you get the same results, you're obligated to agree or propose an alternate theory. Most scientists took the easy way out, however, and simply ignored Burr and Lund. Their discoveries remained little known, and most biologists didn't connect them with the tentative morphogenetic-field concept of regeneration.

Then in 1952 Lund's work was taken up by G. Marsh and H. W. Beams using the planarian. They found that the flatworm's polarity, like the hydra's, could be controlled by passing a current through it. When a direct current was fed in the proper direction through a section of a worm, normal polarity disappeared and a head formed at each end. As the current strength was increased, the section's polarity reversed; a head regrew at the rear, a tail at the front. At higher voltages, even intact worms completely reorganized, with the head becoming a tail and vice versa. Marsh and Beams grew convinced that the animal's electric field *was* the morphogenetic organizing principle. Still, their work was also ignored, except by Meryl Rose, who suggested that a gradation of electrical charge from front to back controlled the gradient of growth inhibitors and stimulators. He suggested that the growth compounds were charged molecules that were moved to different places in the body by the electric field, depending on the amount and sign of their charge and their molecular weight.

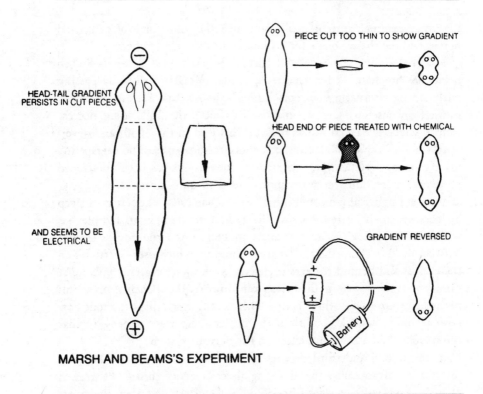

PIECE CUT TOO THIN TO SHOW GRADIENT

HEAD-TAIL GRADIENT PERSISTS IN CUT PIECES

HEAD END OF PIECE TREATED WITH CHEMICAL

AND SEEMS TO BE ELECTRICAL

GRADIENT REVERSED

Battery

MARSH AND BEAMS'S EXPERIMENT

Undercurrents in Neurology

While the investigation of the total body field moved haltingly forward in the study of simple animals, several neurophysiologists began finding out odd things about the nerves of more complex creatures, data that Bernstein's action potential couldn't explain. Going through the old literature, following lead after lead from one paper to the next, I found many hints that there *were* DC potentials in the nervous system and that small currents from outside could affect brain function.

The first recorded use of currents on the nervous system was by Giovanni Aldini, a nephew of Galvani and an ardent champion of vitalism. Using the batteries of his archenemy Volta, Aldini claimed remarkable success in relieving asthma. He also cured a man who today would probably be diagnosed as schizophrenic, although it's impossible to know how much benefit came from the currents and how much from simple solicitude, then so rare in treating mental illness. Aldini gave his patient a room in his own house and later found him a job. Some of Aldini's experiments were grotesque—he tried to resurrect recently executed criminals by making the corpses twitch with electricity—but his idea that external current could replenish the vital force of exhausted nerves became the rationale for a whole century of electrotherapy.

Modern studies of nerves and current began in 1902, when French researcher Stephane Leduc reported putting animals to sleep by passing fairly strong alternating currents through their heads. He even knocked himself unconscious several times by this method. (Talk about dedication to science!) Several others took up this lead in the 1930s and developed the techniques of electroshock and electronarcosis. The therapeutic value of using large currents to produce convulsions has been questioned more and more, until now it's mostly used to quiet unmanageable psychotics and political nonconformists. Electronarcosis—induction of sleep by passing small currents across the head from temple to temple—is widely used by legitimate therapists in France and the Soviet Union. Russian doctors claim their *elektroson* technique, which uses electrodes on the eyelids and behind the ears to deliver weak direct currents pulsing at calmative brain-wave frequencies, can impart the benefits of a full night's sleep in two or three hours. There's still much dispute about how both techniques work, but from the outset there was no denying that the currents had a profound effect on the nervous system.

In the second and third decades of this century there was a flurry of interest in galvanotaxis, the idea that direct currents guided the growth of cells, especially neurons. In 1920, S. Ingvar found that the fibers growing out of nerve cell bodies would align themselves with a nearby flow of current and that the fibers growing toward the negative electrode were different from those growing toward the positive one. Paul Weiss soon "explained" this heretical observation as an artifact caused by stretching of the cell culture substrate due to contact with the electrodes. Even after Marsh and Beams proved Weiss wrong in 1946, it took many more years for the scientific community to accept the fact that neuron fibers *do* orient themselves along a current flow. Today the possible use of electricity to guide nerve growth is one of the most exciting prospects in regeneration research (see Chapter 11).

The Bernstein hypothesis, unable to account for these facts, has turned out to be deficient in several other respects. To begin with, according to the theory, an impulse should travel with equal ease in either direction along the nerve fiber. If the nerve is stimulated in the middle, an impulse should travel in both directions to opposite ends. Instead, impulses travel only in one direction; in experiments they can be made to travel "upstream," but only with great difficulty. This may not seem like such a big deal, but it is very significant. Something seems to polarize the nerve.

Another problem is the fact that, although nerves are essential for regeneration, the action potentials are silent during the process. No impulses have ever been found to be related to regrowth, and neurotransmitters such as acetylcholine have been ruled out as growth stimulators.

In addition, impulses always have the same magnitude and speed. This may not seem like such a big thing either, but think about it. It means the nerve can carry only one message, like the digital computer's 1 or 0. This is okay for simple things like the knee-jerk reflex. When the doctor's rubber hammer taps your knee, it's actually striking the patellar tendon, giving it a quick stretch. This stimulates stretch receptors (nerve cells in the tendon), which fire a signal to the spinal cord saying, "The patellar tendon has suddenly been stretched." These impulses are received by motor (muscle-activating) neurons in the spinal cord, which send impulses to the large muscle on the front of the thigh, telling it to contract and straighten the leg. In everyday life, the reflex keeps you from falling in a heap if an outside force suddenly bends your knees.

The digital impulse system accounts for this perfectly well. However, no one can walk on reflexes alone, as victims of cerebral palsy know all too well. The motor activities we take for granted—getting out of a chair and walking across a room, picking up a cup and drinking coffee, and so on—require *integration* of all the muscles and sensory organs working smoothly together to produce coordinated movements that we don't even have to think about. No one has ever explained how the simple code of impulses can do all that. Even more troublesome are the higher processes, such as sight—in which somehow we interpret a constantly changing scene made of innumerable bits of visual data—or the speech patterns, symbol recognition, and grammar of our languages. Heading the list of riddles is the "mind-brain problem" of consciousness, with its recognition, "I am real; I think; I am something special." Then there are abstract thought, memory, personality, creativity, and dreams. The story goes that Otto Loewi had wrestled with the problem of the synapse for a long time without result, when one night he had a dream in which the entire frog-heart experiment was revealed to him. When he awoke, he knew he'd had the dream, but he'd forgotten the details. The next night he had the same dream. This time he remembered the procedure, went to his lab in the morning, did the experiment, and solved the problem. The inspiration that seemed to banish neural electricity forever can't be explained by the theory it supported! How do you convert simple digital messages into these complex phenomena? Latter-day mechanists have simply postulated brain circuitry so intricate that we will probably never figure it out, but some scientists have said there must be other factors.

Even as Loewi was finishing his work on acetylcholine, others began to find evidence that currents flowed in the nerves. English physiologist Richard Caton had already claimed he'd detected an electric field around the heads of animals in 1875, but it wasn't until 1924 that German psychiatrist Hans Berger proved it by recording the first electroencephalogram (EEG) from platinum wires he inserted into his son's scalp. The EEG provided a record of rhythmic fluctuations in potential voltage over various parts of the head. Berger at first thought there was only one wave from the whole brain, but it soon became clear that the waves differed, depending on where the electrodes were put. Modern EEGs use as many as thirty-two separate channels, all over the head.

The frequency of these brain waves has been crudely correlated with states of consciousness. Delta waves (0.5 to 3 cycles per second) indicate deep sleep. Theta waves (4 to 8 cycles per second) indicate trance, drowsiness, or light sleep. Alpha waves (8 to 14 cycles per second) appear during relaxed wakefulness or meditation. And beta waves (14 to 35 cycles per second), the most uneven forms, accompany all the modulations of our active everyday consciousness. Underlying these rhythms are potentials that vary much more slowly, over periods as long as several minutes. Today's EEG machines are designed to filter them out because they cause the trace to wander and are considered insignificant anyway.

There's still no consensus as to where the EEG voltages come from. They would be most easily explained by direct currents, both steady state and pulsing, throughout the brain, but that has been impossible for most biologists to accept. The main alternative theory, that large numbers of neurons firing simultaneously can mimic real electrical activity, has never been proven.

In 1939, W. E. Burge of the University of Illinois found that the voltage measured between the head and other parts of the body became more negative during physical activity, declined in sleep, and reversed to positive under general anesthesia. At about the same time a group of physiologists and neurologists at Harvard Medical School began studying the brain with a group of MIT mathematicians. This association was destined to change the world. From it came many of our modern concepts of cybernetics, and it became the nucleus of the main American task force on computers in World War II. One of the group's first important ideas was that the brain worked by a combination of analog and digital coding.

One of the mathematicians, computer pioneer John von Neumann, later elaborated the concept in great detail, but basically it's rather simple. In analog computers, changes in information are expressed by analogous changes in the magnitude or polarity of a current. For example, if the computer is to use and store the varying temperatures of a furnace, the rise and fall in heat can be mimicked by a rise and fall in voltage. Analog systems are slow and can handle only simple information, but they can express subtle variations very well. Digital coding, on the other hand, can transmit enormous amounts of data at high speed, but only if the information can be reduced to yes-no, on-off bits—the digits 1 and 0. If the brain was such a hybrid computer, these early cyberneticists reasoned, then analog coding could control the overall activity of large groups of neurons by such actions as increasing or decreasing their sensitivity to incoming messages. (A few years later neurologists did find that some neurons were "tuned" to fire only if they received a certain number of impulses.) The digital system would transfer sensory and motor information, but the processing of that information—memory and recall, thought, and so on—would be accomplished by the synergism of both methods. The voltage changes Burge found in response to major alterations of consciousness seemed to fit within this framework, and his observations were extended by the Harvard-MIT group and others. Much of this work was done directly on the exposed brains of animals and of human patients during surgery. When cooperative patients elected to remain awake during such operations (the brain is immune to pain), human sensations could often be correlated with electrical data. Contributors to this endeavor included nearly all of the greatest American neurophysiologists—Walter B. Cannon, Arturo Rosenblueth, Ralph Gerard, Gilbert Ling, Wilder Penfield, and others.

Measurements on the exposed brain quickly confirmed the existence of potential voltages and also revealed possible currents of injury. Whenever groups of nerve cells were actively conducting impulses, they also produced a negative potential. Positive potentials appeared from injured cells when the brain had been damaged; these potentials then expanded outward to uninjured cells, suppressing their ability to send or receive impulses. When experimenters applied small negative voltages to groups of neurons, their sensitivity increased; that is, they would generate an impulse in response to a weaker stimulus. Externally applied positive potentials worked in the opposite way: They depressed nerve function, making it harder to produce an impulse. Thus there did seem to be an analog code, but how did it work? Did the potentials come from direct currents generated by the nerve cells themselves, or did they merely result from adding up a lot of action potentials all going in the same direction and arriving in the same place at the same time?

Some answers were provided by a series of beautiful experiments by Ling, Gerard, and Benjamin Libet at the University of Chicago. Working on frogs, they studied areas of the cortex where the neuron layer was only one cell thick and the cells were arranged side by side like soldiers on review, all pointing in the same direction. In such areas they found a negative potential on the dendrites (the short incoming fibers) and a positive potential at the ends of the axons (the longer outgoing fibers). This indicated a steady direct current along the normal direction of impulse transmission. The entire nerve cell was electrically polarized.

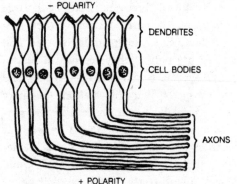

POLARIZED NERVE CELLS LINED UP IN CEREBRAL CORTEX

In another series of experiments, on brains removed from frogs and kept alive in culture, the Chicago group found that direct currents swept across the surface of the cortex in very slow waves, which could be produced experimentally by applying chemicals such as caffeine to a single spot on the surface. When they made a cut on the brain, severing groups of nerve fibers, these DC traveling waves would still cross the cut *if the two surfaces were in direct contact.* If the researchers held the cut open and filled it with a saline solution that matched body fluids, then the waves couldn't cross the gap. These were particularly important observations. They indicated that the current was transmitted by structures *outside* the neurons; it crossed the cut when the edges touched, but the microscopic parts of the severed neurons wouldn't have rejoined so easily. The results also showed that the current was *not* a flow of ions; otherwise it would have been able to cross the gap through the salt water.

Studying intact brains in living frogs, the same group found a potential between the front and back of the brain. The olfactory (frontal) lobes were several millivolts negative with respect to the occipital (rear) lobe, implying a current flowing up the brain stem and between the two hemispheres to the front.

At the time, these observations seemed mighty odd. They didn't fit any concepts of how the nerves worked. As a result, they were largely ignored. The majority of neurophysiologists went on measuring the action potentials and tracing out fiber pathways in the brain. This was useful work but limited. The basic questions remained.

Only one research team followed up this work, some ten years later. Sidney Goldring and James L. O'Leary, neuropsychiatrists at the Washington University School of Medicine in St. Louis, recorded the same DC potentials from the human scalp, from the exposed brain during surgery, and from the brains of monkeys and rabbits. As noted before, the potentials varied in regular cycles several minutes long, like a basso continuo under the EEG. In fact, Goldring and O'Leary found waves within waves: "Written upon the slow major swings were lesser voltage changes." These were weak potentials, measured in microvolts (millionths of a volt) and varying in waves of 2 to 30 cycles per minute, sort of a pianissimo "inner voice" in a three-part electrical fugue.

Conducting in a New Mode

I was acutely aware that I didn't have the "proper" background for the work I planned to do. I wasn't a professional neurophysiologist; I didn't even know one. Indeed, after my run-in with the research committee, one member had taken me aside and earnestly advised me, "Go back to school and get your Ph.D., Becker. Then you'll learn all of this stuff is nonsense." Still, some of the greatest neurophysiologists had thought the same way I did about "all of this stuff." They suggested we might have been too hasty in throwing electric currents out of biology. My notion of putting them back in wasn't so outlandish, but only an extension of what they'd been saying. I was approaching the body's system of information transfer from the periphery, asking, "What makes wounds heal?" They'd started from the center, asking, "How does the brain work?" We were working on the same problem from opposite ends. As I contemplated their findings and all of biology's unsolved problems, I grew convinced that life was more complex than we suspected. I felt that those who reduced life to a mechanical interaction of molecules were living in a cold, gray, dead world, which, despite its drabness, was a fantasy. I didn't think electricity would turn out to be any *élan vital* in the old sense, but I had a hunch it would be closer to the secret than the smells of the biochemistry lab or the dissecting room's preserved organs.

I had another worthy ally when I started to reevaluate the role of electricity in life. Albert Szent-Györgyi, who'd already won a Nobel Prize for his work on oxidation and vitamin C, made a stunning suggestion in a speech before the Budapest Academy of Science on March 21, 1941. (Think of the date. World War II was literally exploding around him, and there he was, calmly laying the foundations for a new biology.) Speaking of the mechanistic approach of biochemistry, he pointed out that when experimenters broke living things down into their constituent parts, somewhere along the line life slipped through their fingers and they found themselves working with dead matter. He said, "It looks as if some basic fact about life were still missing, without which any real understanding is impossible." For the missing basic fact, Szent-Györgyi proposed putting electricity back into living things, but not in the way it had been thought of at the turn of the century.

At that earlier time, there had been only two known modes of current conduction, ionic and metallic. Metallic conduction can be visualized as a cloud of electrons moving along the surface of metal, usually a wire. It can be automatically excluded from living creatures because no one has ever found any wires in them. Ionic current is conducted in solutions by the movement of ions—atoms or molecules charged by having more or fewer than the number of electrons needed to balance their protons' positive charges. Since ions are much bigger than electrons, they move more laboriously through the conducting medium, and ionic currents die out after short distances. They work fine across the thin membrane of the nerve fiber, but it would be impossible to sustain an ionic current down the length of even the shortest nerve.

Semiconduction, the third kind of current, was a laboratory curiosity in the 1930s. Halfway between conductors and insulators, the semiconductors are inefficient, in the sense that they can carry only small currents, but they can conduct their currents readily over long distances. Without them, modern computers, satellites, and all the rest of our solid-state electronics would be impossible.

Semiconduction occurs only in materials having an orderly molecular structure, such as crystals, in which electrons can move easily from the electron cloud around one atomic nucleus to the cloud around another. The atoms in a crystal are arranged in neat geometrical lattices, rather than the frozen jumble of ordinary solids. Some crystalline materials have spaces in the lattice where other atoms can fit. The atoms of these impurities may have more or fewer electrons than the atoms of the lattice material. Since the forces of the latticework structure hold the same number of electrons in place around each atom, the "extra" electrons of the impurity atoms are free to move through the lattice without being bound to any particular atom. If the impurity atoms have fewer electrons than the others, the "holes" in their electron clouds can be filled by electrons from other atoms, leaving holes elsewhere. A negative current, or N-type semiconduction, amounts to the movement of excess electrons; a positive current, or P-type semiconduction, is the movement of these holes, which can be thought of as positive charges.

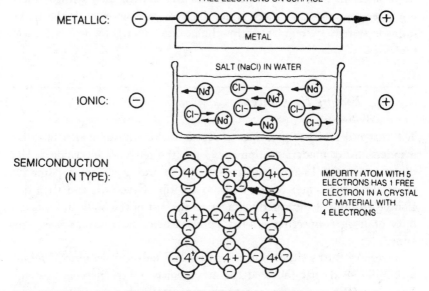

THREE WAYS OF CONDUCTING ELECTRICAL CURRENT

Szent-Györgyi pointed out that the molecular structure of many parts of the cell was regular enough to support semiconduction. This idea was almost completely ignored at the time. Even when Szent-Györgyi expanded the concept in his 1960 *Introduction to a Submolecular Biology*, most scientists (except in Russia!) dismissed it as evidence of his advancing age, but that little book was an inspiration to me. I think it may turn out to be the man's most important contribution to science. In it he conjectured that protein molecules, each having a sort of slot or way station for mobile electrons, might be joined together in long chains so that electrons could flow in a semiconducting current over long distances without losing energy, much as in a game of checkers one counter could jump along a row of other pieces across the entire board. Szent-Györgyi suggested that the electron flow would be similar to photosynthesis, another process he helped elucidate, in which a kind of waterfall of electrons cascaded step by step down a staircase of molecules, losing energy with each bounce. The main difference was that in protein semiconduction the electrons' energy would be conserved and passed along as information instead of being absorbed and stored in the chemical bonds of food.

With Szent-Györgyi's suggestion in mind, I put together my working hypothesis. I postulated a primitive, analog-coded information system that was closely related to the nerves but not necessarily located in the nerve fibers themselves. I theorized that this system used semiconducting direct currents and that, either alone or in concert with the nerve impulse system, it regulated growth, healing, and perhaps other basic processes.

Testing the Concept

The first order of business was to repeat Burr's measurements on salamanders, using modern equipment. I put the reference electrode at the tip of each animal's nose and moved the recording electrode point by point along the center of the body to the tip of the tail, and then out along each limb. I measured voltages on the rest of the body and plotted lines of force connecting all the points where the readings were the same.

Instead of Burr's simple head-negative and tail-positive form, I found a complex field that followed the arrangement of the nervous system. There were large positive potentials over each lobe of the brain, and slightly smaller ones over the brachial and lumbar nerve ganglia between each pair of limbs. The readings grew increasingly negative as I moved away from these collections of nerve cell bodies; the hands, feet, and tip of the tail had the highest negative potentials.

In another series of measurements, I watched the potentials develop along with the nervous system in larval salamanders. In the adults, cutting the nerves where they entered the legs—that is, severing the long nerve fibers from their cell bodies in the spinal cord—wiped out the limb potentials almost entirely. But if I cut the spinal cord, leaving the peripheral nerves connected to their cell bodies, the limb potentials didn't change. It certainly looked as though there was a current being generated in the nerve cell bodies and traveling down the fibers.

To have a current flow you need a circuit; the current has to be made at one spot, pass through a conductor, and eventually get back to the generator. We tend to forget that the 60-cycle alternating current in the

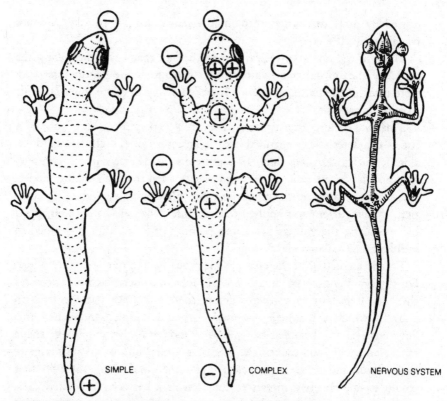

SIMPLE   COMPLEX   NERVOUS SYSTEM

ELECTRICAL MEASUREMENTS ON THE SALAMANDER

wall socket isn't used up when we turn on a light but is merely coursing through it to the ground, through which it eventually returns to the power station. Since my measurements were positive over collections of nerve cell bodies, and increasingly negative out along the nerve fibers, it seemed a good bet that current was being generated in the cell bodies, especially since they contained all of the "good stuff"—the nucleus, organelles, and metabolic components—while the fibers were relatively uninteresting prolongations of the body. At the time, I supposed the circuit was completed by current going back toward the spine through the muscles.

This was a good start, but it wasn't scientifically acceptable proof. For one thing, my guess about the return part of the circuit was soon disproved when I measured the limb muscles and found them polarized in the *same* direction as the surface potentials. For another thing, it had recently been discovered that amphibian skin itself was polarized, inside versus outside, by ion differences much like the nerve membrane's resting potential, so it was just barely possible that my readings had been caused by ionic discharges through the moist skin. If so, my evidence was literally all wet.

Much of the uncertainty was due to the fact that I was measuring the *outside* of the animal and assuming that generators and conductors *inside* were making the pattern I found. I needed a way to relate inner currents to outer potentials.

This was before transistors had entirely replaced vacuum tubes. A tube's characteristics depended on the structure of the electric field inside it, but to calculate the field parameters in advance without computers was a laborious task, so radio engineers often made an analog model. They built a large mock-up of the tube, filled with a conducting solution. When current was applied to the model, the field could be mapped by measuring the voltage at various points in the solution. I decided to build a model salamander.

I made an analog of the creature's nervous system out of copper wires. For the brain and nerve ganglia I used blobs of solder. Each junction was thus a voltaic battery of two different metals, copper and the lead-tin alloy of which the solder was made. Then I simply sandwiched this "nervous system" between two pieces of sponge rubber cut in the shape of a salamander, and soaked the model in a salt solution to approximate body fluids and serve as the electrolyte, the conducting solution that would enable the two metals to function as a battery. It worked. The readings were almost exactly the same as in the real salamander. This showed that a direct current inside *could* produce the potentials I was getting on the outside.

If my proposed system was really a primitive part of the nervous system, it should be widely distributed, so next I surveyed the whole animal kingdom. I tested flatworms, earthworms, fish, amphibians, reptiles, mammals, and humans. In each species the potentials on the skin reflected the arrangement of the nervous system. In the worms and fish, there was only one area of positive potential, just as there was only one major nerve ganglion, the brain. In humans the entire head and spinal region, with its massive concentration of neurons, was strongly positive. The three specific areas of greatest positive potential were the same as in the salamander: the brain, the brachial plexus between the shoulder blades, and the lumbar enlargement at the base of the spinal cord. In all vertebrates I also recorded a midline head potential that suggested a direct current like that postulated by Gerard, flowing from back to front through the middle of the brain. It looked as though the current came from the reticular activating system, a network of cross-linked neurons that fanned out from the brainstem into higher centers and seemed to control the level of sleep or wakefulness and the focus of attention.

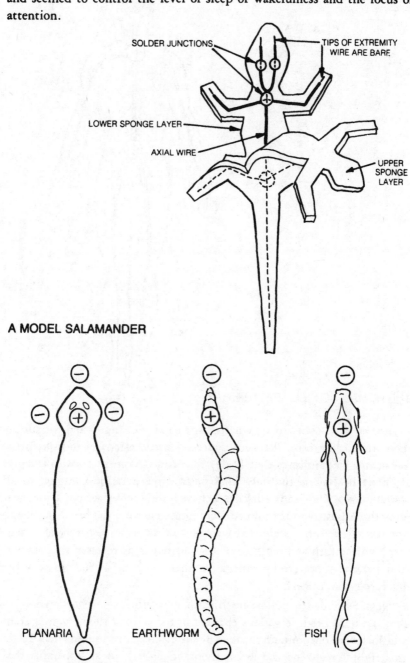

A MODEL SALAMANDER

A DC FIELD PATTERN APPEARS IN ALL ANIMALS

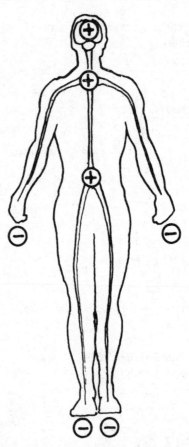

THE HUMAN DC FIELD PATTERN

At the same time, to see whether the current of injury and the surface potentials came from the same source, I made electrical measurements on salamander limbs as they healed fractures. (As mentioned in Chapter 1, bone healing is the only kind of true regeneration common to all vertebrates.) The limb currents behaved like those accompanying regrowth. A positive zone immediately formed around the break, although the rest of the limb retained at least part of its negative potential. Then, between the fifth and tenth days, the positive zone reversed its potential and became more strongly negative than the rest of the limb as the fracture began to heal.

Next I decided to follow up Burge's experiments of two decades before. I would produce various changes in the state of the nervous system and look for concomitant changes in the electrical measurements. To do this right I really needed a few thousand dollars for an apparatus that could take readings from several electrodes simultaneously and record them side by side on a chart. My chances of getting this money seemed

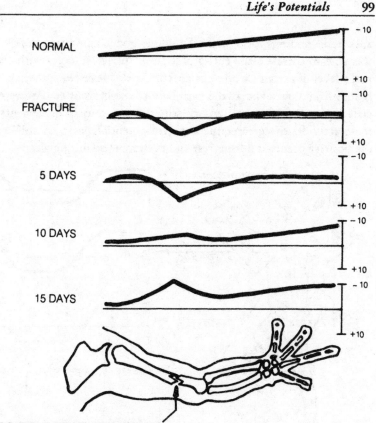

THE DC FIELD NEAR A FRACTURE

slim unless I could publish another paper fast. I decided to use the equipment I had for a simple measurement during one of the most profound changes in consciousness—anesthesia.

Burge was right. The electrical responses were dramatic and incontrovertible. As each animal went under, its peripheral voltages dropped to zero, and in very deep anesthesia they reversed to some extent, the limbs and tail going positive. They reverted to normal just before the animal woke up.

I had enough for a short paper, and I decided to try a journal on medical electronics recently started by the Institute of Radio Engineers. Although most of what they printed was safe and unremarkable, I'd found that engineers were often more open-minded than biologists, so I went for broke; I put in the whole hypothesis—analog nervous system, semiconducting currents, healing control, the works. The editor loved it and sent me an enthusiastic letter of acceptance, along with suggestions for further research. Best of all, I soon got another small grant approved and bought my multi-electrode chart recorder. Soon I had confirmed my anesthesia findings, and with the whole-body monitoring setup I also was able to correlate the entire pattern of surface voltages with the animal's level of activity while not anesthetized. Negative potentials in the brain's frontal area and at the periphery of the nervous system were associated with wakefulness, sensory stimuli, and muscle movements. The more activity, the greater the negative potentials were. A shift toward the positive occurred during rest and even more so during sleep.*

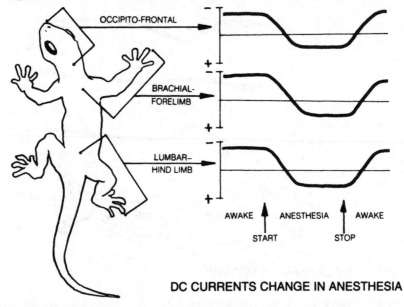

OCCIPITO-FRONTAL

BRACHIAL-
FORELIMB

LUMBAR–
HIND LIMB

AWAKE   ANESTHESIA   AWAKE
START   STOP

DC CURRENTS CHANGE IN ANESTHESIA

In my reading on solid-state electronics I found another way I could test for current in the salamander. Luckily it was cheap and easy; I could do it without buying more equipment. Best of all, it should work only if the current was semiconducting.

Suppose you think you have a current flowing through some conductor—a salamander's limb, for instance. You put it in a strong magnetic field so that the lines of force cut across the conductor at right angles. Then you place another conductor, containing no current, perpendicular to both the original conductor (the limb) and the magnetic field. If there is a current in whatever you're testing, some of the charge carriers will be deflected by the magnetic field into the other conductor, producing a voltage that you can measure. This is called the Hall voltage, after the gentleman who discovered it. The beauty of it is that it works differently for the three kinds of current. For any given strength of mag-

*I didn't know it until later, but another experimenter named H. Caspers made similar findings at about the same time.

netic field, the Hall voltage is proportional to the mobility of the charge carriers. Ions in a solution are relatively big and barely deflected by the field. Electrons in a wire are constrained by the nature of the metal. In both cases the Hall voltage is small and hard to detect. Electrons in semiconductors are very free to move, however, and produce Hall voltages with much weaker fields.

After finding a C-shaped permanent magnet, an item not much in demand since the advent of electromagnets, I set up the equipment. I took a deep breath as I placed the first anesthetized salamander on its plastic support, with one foreleg extended. I'd placed electrodes so that they touched the limb lightly, one on each side, and I'd mounted the magnet so as to swing in with its poles above and below the limb, close to yet not touching it. I took voltage measurements every few minutes, with the magnet and without it, as the animal regained consciousness. I also measured the DC voltage from the tips of the fingers to the spinal cord. In deep anesthesia, the DC voltage along the limb was zero and so was the Hall voltage. As the anesthetic wore off, the normal potential along the limb gradually appeared, and so did a beautiful Hall voltage. It increased right along with the limb potential, until the animal recovered completely and walked away from the apparatus. The test worked every time, but I don't think I'll ever forget the thrill of watching the pen on the recorder trace out the first of those Hall voltages.

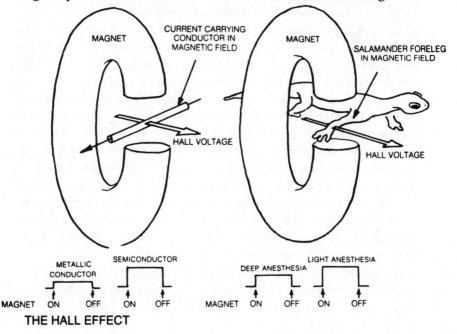

THE HALL EFFECT

This experiment demonstrated unequivocally that there was a real electric current flowing along the salamander's foreleg, and it virtually proved that the current was semiconducting. In fact, the half-dozen tests I'd performed supported every point of my hypothesis.

Scientific results that aren't reported might as well not exist. They're like the sound of one hand clapping. For scientists, communication isn't only a responsibility, it's our chief pleasure. A good result from a clean, beautiful experiment is a joy that you just *have* to share, and I couldn't wait to see these data in print. I went for the top this time. *The* journal in American science is aptly named *Science.* Each issue reports on all fields from astronomy to zoology, so publication means a paper has more than a specialized significance. Mine was accepted, and I was jubilant.

With three major papers in three major journals after my first year of research, I felt I'd arrived. The world has a way of cutting you down to size, however, and in the science game the method is known as citation. No matter how important your paper is, it doesn't mean anything unless it's cited as a reference in new papers by others and you get a respectable number of requests for reprints. On both counts, I was a failure. I was learning how science treats new ideas that conflict with old ones.

I didn't stay discouraged long, though. I was doing science for the love of it, not for praise. I felt the concepts emerging from my reading and research were important, and I was passionately committed to testing them. I knew that if the results were ever to change any minds, I would have to be careful not to misinterpret data. In going deeper into the electrical properties of nerves, I realized, I was about to get over my head in an area I *really* wasn't trained for—physics. I made one of the best decisions of my life; I looked for a collaborator.

The basic scientists at the State University of New York Upstate Medical Center, the medical school affiliated with the VA hospital, were not only uninterested, they were horrified at what I was doing and wouldn't risk their reputations by becoming associated with me in any way. So, I walked across the street to the physics department of Syracuse University and spoke to the chairman, an astronomer whom I'd met a few years earlier when I volunteered to watch the northern lights during the International Geophysical Year. After a few minutes' thought, he suggested that a guy on the third floor named Charlie Bachman might be "as crazy as you," and wished me luck.

The instant I opened the door, I knew I was in the right place. There was Charlie, bent over a workbench with an electromagnet and a live frog.

Five

The Circuit of Awareness

Charlie and I talked all afternoon, beginning fifteen years of fruitful work together. For me, the best part was his friendship and his open mind. He, too, knew there was still a lot to learn. Our relationship also had a side effect of incalculable value: He sent some of his most talented graduate students over to my lab to do their thesis work and later to become my colleagues in research. Andy Marino, Joe Spadaro, and Maria Reichmanis each became an indispensable part of the research group. Like Charlie, they constantly contributed new ideas, and they helped create the atmosphere of intellectual adventurousness that makes a lab creative.

Closing the Circle

Charlie's first contribution was to check the equipment and confirm the measurements I'd made on the salamanders. After he'd satisfied himself that everything was real, we discussed what to do next.

"Well," Charlie said, "to find out more about this current, we'll have to go into the animal—expose a nerve and measure the current there."

"That's easier said than done," I objected. "Just to cut down into the leg of an animal will damage tissues and produce currents of injury. That'll give spurious voltages. Besides, there'd be no stable place to put the reference electrode."

Charlie gave me a lesson in basic electronics. A voltage entering a wire will decrease as the current travels along, so there'll be a uniform voltage drop in each unit of length. All you need to do is put both electrodes along the conductor, with the reference electrode closer to what you think is the source. If you use a standard distance between the electrodes, you can compare the voltage drop along various wires and measure changes in the whole system from any segment of it.

All I had to do was the surgery. I decided to work on bullfrogs, whose hind legs were long and contained a nice big sciatic nerve. It was easy to find and could be exposed with just a little careful dissection, going between muscles instead of cutting through them. I was able to isolate over an inch of nerve with no bleeding or tissue damage, slipping a plastic sheet underneath so as not to pick up readings from the surrounding muscles. We measured the voltage gradient over a standard distance of 1 centimeter. It was the same from one frog to the next. In deep anesthesia it was absent or pretty small; as the anesthesia wore off, it became a constant drop of about 4 millivolts per centimeter, always gradually positive toward the spinal cord and negative toward the toes.

In some frogs we cut the nerve above the measuring site, whereupon the voltages disappeared—another indication that the current was actually in the nerve. Voltages returned a little later, but they weren't the same as before. We figured these secondary voltages were probably an artifact—a spurious measurement produced by extraneous factors—caused by currents of injury from the cut nerve itself or from the other tissues where I'd made the incision to cut the nerve.

Charlie then suggested that we make measurements on a longer section of nerve, and that was when we ran into a puzzle. The nicely reproducible voltages we'd found before couldn't be duplicated when we extended our measurement distance to 2 centimeters, close to the knee. We expected double the potential we'd found over the 1-centimeter distance, but often it was lower or higher than it should have been. I insisted that my dissection was producing local currents of injury that made our readings unpredictable. However, Charlie pointed out that I was a good frog surgeon and I didn't seem to be doing any more damage than before. He asked, "Could there be any difference in the nerve where you extended the dissection?"

"Not likely," I said. "The sciatic nerve does split up into two branches, but you only find them below the knee, where one goes to the front of the calf and one to the back."

"How do you know it doesn't separate before it gets to the knee?" he asked.

He was right. Not bad for a physicist! The nerve did divide, but the two parts were held together by the nerve sheath until they got past the knee. I was able to remove the sheath and isolate both portions. When we measured these, we found that the two sections were polarized in opposite directions. The voltage drop of the front branch was positive toward the toes. The posterior branch was polarized in the same direction as the sciatic trunk, but it always had a higher voltage gradient. The current in the front branch apparently flowed in the direction opposite to that in the rest of the nerve. The interesting thing was, when we added up the voltage increases from 1-centimeter lengths of the two branches—4 millivolts positive and 8 negative in a typical frog—we got roughly the same voltage gradient that we found in the main nerve, about 4 millivolts negative in every centimeter. At first that didn't make any sense.

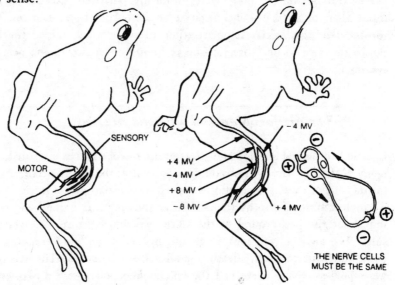

FRONT BRANCH: MAINLY SENSORY

REAR BRANCH: MAINLY MOTOR

BRANCHES ARE ELECTRICALLY POLARIZED IN OPPOSITE DIRECTIONS

THE NERVE CELLS MUST BE THE SAME

NERVES MUST HAVE A LONGITUDINAL CURRENT

THE DC CIRCUIT IN FROGS

On a hunch I took pieces of each nerve and sent them to the pathology department to have microscope slides made. I found that the fibers in the front fork were smaller than those in the other. A light bulb went on! The sciatic nerve is what's called a mixed nerve. It has both motor and sensory neurons. Sensory fibers are usually narrower than motor fibers, so it looked as though the front branch was all sensory, the back one all motor. Suppose the DC system also had incoming and outgoing divisions. We took readings from other nerves known to be all one type or the other. The femoral nerve along the front of the thigh is almost entirely motor in function, and, sure enough, it had an increasing negative potential *away* from the spine. The spinal nerves that serve the skin of a frog's back are sensory fibers, and they had increasing negative voltages *toward* the spine.

Now we saw that when you put motor and sensory nerves together into a reflex arc, the current flow formed an unbroken loop. This solved the mystery of what completed the circuit: The current returned through nerves, not some other tissue. Just as Gerard had found in the brain, nerves throughout the body were uniformly polarized, positive at the input fiber, or dendrite, and negative at the output fiber, or axon. We realized that this electrical polarization might be what guided the impulses to move in one direction only, giving coherence to the nervous system.

The Artifact Man and a Friend in Deed

Charlie had helped develop the electron microscope and as a result knew many of the big names in physiology. Soon after the sciatic nerve experiments, one of these acquaintances visited Syracuse to give a lecture, and we invited him to stop by the lab. After showing him around and talking about the background of the work, we showed him our latest results. We anesthetized four frogs and opened their legs, exposing all eight sciatic nerves and measuring all sixteen branches. The readings were flawless. Every nerve had the voltage and polarity we'd predicted. Proudly, we asked, "What do you think?"

"Artifact, all artifact," he replied. "Everyone knows there's no current along the nerve." Just then he remembered he had pressing business elsewhere and left in a hurry, apparently afraid some of this might rub off on him.

Charlie almost never swore, but that day he did. The gist of his remarks was that there sure was a difference between physicists and biologists. The former would at least look at new evidence, while the latter kept their eyes and minds closed. Thereafter we always referred to the "Artifact Man" when we needed a symbol of dogmatism.

We continued a few more observations on frog nerves. By now winter had come. That shouldn't have mattered—the lab temperature was the same all year, and the frogs didn't stop eating and hibernate as they would have in the wild—but there *was* a difference. The frogs' voltages were much lower, they stayed unconscious longer with the standard dose of anesthetic, and their blood vessels were much more fragile. Did they somehow sense the winter?

If the DC system was as we theorized, it would be influenced by external magnetic fields. In the Hall-effect experiment I'd already shown that it was, but I'd used a strong field, measuring thousands of gauss. The earth's magnetic field is only about half a gauss, but it does vary in a yearly cycle. At the time there was another scientist who was saying this weak field had major effects on all life. Frank A. Brown, a Northwestern University biologist who was studying the ubiquitous phenomenon of biological cycles—wavelike changes in metabolic functions, such as the alternation of sleep and wakefulness—was claiming that similar rhythms in the earth's magnetic field served as timers for the rhythms of life. Even though his evidence was good, no one paid any attention to him in the early sixties, but it seemed to me that we had something to offer Brown's effort. We had a link by means of which the effect could occur.

I wrote up the sciatic nerve measurements and added the observation on winter frogs. I sent it to *Science* but got it back immediately. I guess the editors had second thoughts after running my paper on the Hall effect. Next I tried the even better British equivalent, *Nature,* which took it. This time I also got some reprint requests. More important, the report led to correspondence with Frank Brown, beginning years of mutual feedback that helped bring about the discoveries described in Chapter 14.

I thought of one more way we could check whether the current in the nerves was semiconducting. We could freeze a section of nerve between the electrodes. If the current was carried by ions, they would be frozen in place and the voltage would drop to zero. However, if the charge carriers were electrons in some sort of semiconducting lattice, their mobility would be *enhanced* by freezing and the voltage would rise.

It worked. Each time I touched the nerve with a small glass tube filled with liquid nitrogen, the voltage shot upward. But perhaps I was damaging the nerve with the glass tube or through the freezing itself. Maybe the increase was merely a current of injury. To check, we simply cut the nerve near the spinal cord; the voltage gradient on the nerve went to zero, and then we applied the liquid nitrogen again. If the cold was really enhancing a semiconducting current, we should find no voltage now even after freezing the nerve—and we didn't. Therefore the increase in current wasn't due to artifact—damage to the nerve by freezing or touching it with the tube.

That settled it. Test after test had substantiated the direct-current system. Now we had to see where the concept would lead us and try to convince some of the Artifact Men along the way. We had lots of ideas for further work, but now the first priority was to get some reliable system of funding for ourselves.

I was continuing to have problems with the VA research office. After I'd gotten my second grant from that source, I soon found out that to have it approved and to be able to spend it were two different things. To order supplies—even things as simple as test tubes or electrode wire—I had to fill out a form and give it to the secretary of the research office. She had to fill out another form and get it signed by the research director. This form went to the supply service, where a clerk filled out a third form to actually order the stuff. Well, my orders stopped getting filled.

In the process of complaining I made friends with the secretaries and found out that the director was holding me up just by not signing my forms. His secretary solved my problem. The director was a procrastinator. A pile of papers would collect on his desk until his secretary told him they had to be taken care of right away. Then he would sign them all at once without looking at each one. His secretary, to whom I owe a tremendous debt, merely slipped my requests back into the middle of the pile, usually late on Friday afternoon. Several times he visited my lab, saw a new piece of equipment, and remarked, "I don't remember ordering that for you."

"You don't?" I replied sweetly. "We talked about it, and I had plenty of money left in the grant, so you said okay." It was better than arguing over each instrument, and I was careful not to overspend. I don't think he ever caught on.

Soon I encountered a more serious threat, however. Between the VA and the medical school I had a lot of bosses, and all of them were doing "research." However, the research service's annual report showed that I'd published more than all of my superiors put together on a few thousand bucks a year, while some of them were drawing forty or fifty thousand. I'd broken the old rule that you should never do more than your boss.

One of these fellows appeared in my lab one day. That was an event in itself, since he'd never been a supporter of mine; in fact, our relations were rather strained. That day, however, he evinced great interest in what I was doing and made me an "offer I couldn't refuse."

"How would you like to have as much money as you need?"

I said that would be nice, but I wondered how it could come about.

"No problem. All you have to do is include me in the project. All I would expect in return is that my name would go on all publications."

It was a few seconds before I could believe I'd heard him right. Then I told him what he could do with his influence.

A few months later, I found out that the area surgical consultant, practically next to God in the VA hierarchy, was visiting the hospital to act on a report, made by my would-be "benefactor," that I was spending too much time on research and neglecting my patients. Fortunately, there was a lot of infighting among my superiors, and one higher than the guy who'd made the charge was supporting me. His motives were less to save a promising research program and more to embarrass the other man, but I was cleared.

It was also clear that I was courting disaster by relying on VA money. I needed outside support. I took time off from research to prepare two proposals. One, which I sent to the Department of the Army, emphasized the possibility that direct currents could stimulate healing. Since the Army's business produces quite a few wounds, I thought it would be interested, but it was not. The proposal was turned down promptly, but then a strange thing happened about a month later. I received a long-distance call from a prominent orthopedic surgeon, a professor at a medical school in the South. "I have a grant from the Army to study the possibility that direct currents might stimulate wound healing," he purred, "and I wonder if you might have any suggestions as to the best approach to use."

My God, were they all this sleazy out there? Of course, when I looked up his credentials, I found he had absolutely no background in bioelectricity. He'd just happened to be on the Army review committee, recommended disapproval, and then turned around and submitted the idea in his own name, now getting the go-ahead since *he,* a man with a reputation and friends on the review board, was going to do it, instead of some unknown upstart.

I sent the second application to the National Institutes of Health (NIH). I stayed within my specialty and proposed to study the solid-state physics of bone, eventually hoping to find out if direct currents could stimulate bone healing. The grant was approved, but only for enough money to do part of what I wanted. And although it was nice to have a cushion, a source not under local control, I nevertheless needed some political clout to stabilize the situation in Syracuse. I went directly to the dean of the medical school.

Carlyle Jacobsen had seemed to be a nice guy, not the type to stand on ceremony or position, and I thought I could talk to him frankly. I gathered up reprints of my papers and went to his office.

"Sir," I began, "I've been doing research on direct current electrical effects in living things for the past four years. I've gotten some papers published in good journals, and I think this is an important piece of work. Nevertheless, I have great difficulty getting funds from the VA. My requests are blocked by the politicians on the committee. Meanwhile these same guys are spending five times as much as I get, and they don't publish a damn thing." I'm afraid I got carried away, but Dean Jacobsen just sat there listening until I'd finished.

"Have you done any experiments on the DC activity of the nervous system?" he asked.

This was an unexpected question, but I told him of our work on salamanders and frog nerves.

It turned out he'd done some research on nerves years ago—with Ralph Gerard! He was very enthusiastic. "You've gone much further than we ever did," he told me. "We never thought to relate the brain currents to a total-body system. How much do you need?"

I asked for $25,000 in each of the next two years, but explained that it had to be earmarked for me alone or I would never see it.

"Don't worry," he said. "Go right back to your lab. I'll get it for you. I wish I could work with you."

He must have been dialing Washington as the door closed behind me, for the next day the chief of research got a telegram from the VA Central Office authorizing the requested amount for me, and only me. He couldn't understand it, and I professed complete ignorance, too.

I figured nothing I did now could make the research director like me any less, so I made another move. I went to the hospital director and told him I needed more space. Having heard of my favor from Washington, he was most helpful, and soon I had a suite of rooms on the top floor.

Suddenly a whole new realm of research was within reach. Charlie and I hardly knew which way to turn. Our first and most important step was to hire Andy Marino as a technician. The salary meant much to him, and his intelligence and dedication meant even more to us. We were on our way.

The Electromagnetic Brain

If the current controlled the way nerves worked in the brain as well as in the rest of the body, then it must regulate consciousness to some extent.

Certainly the falling voltages in anesthetized salamanders supported this idea. The question was: Did the change in the current *produce* anesthesia? Apparently it did, for when I passed a minute current front to back through a salamander's head so as to cancel out its internal current, it fell unconscious. How this state compared with normal sleep was impossible to tell, but at least the animal was clinically anesthetized. As long as the current was on, the salamander was motionless and unresponsive to painful stimuli.

Was this real anesthesia, or was the animal just being continuously shocked? This certainly didn't seem to be the case, but the observation was so important and so basic to neurophysiology that I had to be sure. It was no easy task, however, for there were, and still are, few objective tests known for anesthesia, especially in salamanders. Brain waves had turned out to be useless in gauging depth of anesthesia in humans, because the one good marker—very slow delta waves—only showed up when the patient was dangerously close to death. Lacking any better idea, however, I used my new multi-electrode monitor to make EEG recordings of chemically anesthetized salamanders and found that they showed prominent delta waves even though they all recovered nicely. Delta waves would be my marker. The idea worked beautifully. Very small currents gave me delta waves on the EEG, the waves got bigger as I increased the current, and they all correlated with the animal's periods of unresponsiveness.

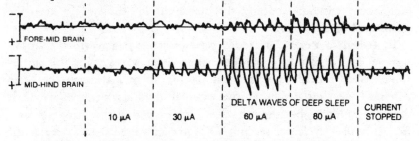

DC THROUGH THE HEAD ANESTHETIZES THE SALAMANDER

This result naturally led to the question: Did chemical anesthetics work by stopping the brain's electrical current? I couldn't see any way to get direct evidence one way or the other, but I thought maybe chemical anesthesia could be reversed by putting current into the brain in the normal direction. I found this could be done only to a certain extent. I could get a partial return of the higher-frequency waves in the EEG, and the anesthesia seemed to become shallower, but I couldn't get a drugged salamander to wake up and walk away.

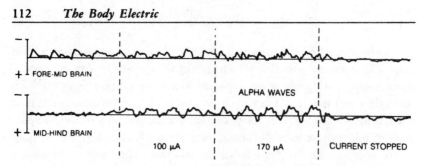

FORE-MID BRAIN

ALPHA WAVES

MID-HIND BRAIN

100 μA   170 μA   CURRENT STOPPED

DC THROUGH THE HEAD PARTIALLY AWAKENS THE SALAMANDER

In the course of these observations, I found that when the head voltage was dropping as a chemical anesthetic took hold, specific slow waves always appeared briefly in the recordings. They were at the low end of the delta frequencies, 1 cycle per second or even less, and they also showed up when the voltage came back as the drug wore off. To find out if these waves always signaled a major change in the state of consciousness, I decided to use a standard amount of direct current to produce anesthesia, measure the amplitude (size) of the delta waves in the EEG, and then add some one-second waves of my own to the current I was putting into the animal's head. In other words, I would introduce some "change-of-state" waves from outside and see if they produced a shift in the EEG. I couldn't record the EEG simultaneously, because the waves I added would appear on the trace, so I rigged up a switch to cut out the added waves after a minute and turn on the EEG recorder at the same time, without stopping the direct current that would keep the salamander unconscious.

It seemed to work. The added waves markedly increased the amplitude of the salamander's own deep-sleep delta waves. Was this an artifact? Were the added waves just causing an oscillation in the brain currents that persisted after the external rhythms were removed? It didn't seem likely, because the waves I added were at the change-of-state frequency of 1 cycle per second, while the measured deltas were at the deep-sleep frequency of 3. However, an additional test was possible. I could add waves of other frequencies and see if they worked as well as 1 cycle per second. They didn't; in fact, as the frequency of the added waves increased over that rate, the deep-sleep delta waves got smaller. The one-second waves *were* a marker of major shifts in consciousness.

This line of work corroborated one of the main points of my hypothesis. Direct currents within the central nervous system regulated the level of sensitivity of the neurons by several methods: by changing the amount of current in one direction, by changing the direction of the current (reversing the polarity), and by modulating the current with slow waves. Moreover, we could exert the same control from outside by putting current of each type into the head. This was exciting. It opened up vast new possibilities for a better understanding of the brain. It was still on the edge of respectability, too, since it was a logical consequence of the work done by Gerard and his co-workers. The next experiment was harder to believe, however.

I figured the brain currents must be semiconducting, like those in the peripheral nerves. I thought of looking for a Hall voltage from the head but reasoned the brain's complexity would make any results questionable. Then I thought of using the effect backward, so to speak, measuring a magnetic field's action on the brain rather than on the production of the Hall voltage. Since the Hall voltage was produced by diverting some of the charge carriers from the original current direction, a strong enough magnetic field should divert all of them. If so, such a field perpendicular to the brain's midline current should have the same effect as canceling out that normal current with one applied from the outside. The animal should fall asleep.

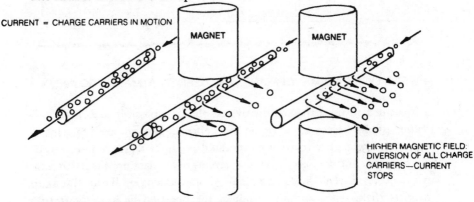

CURRENT = CHARGE CARRIERS IN MOTION

MAGNET   MAGNET

HIGHER MAGNETIC FIELD: DIVERSION OF ALL CHARGE CARRIERS—CURRENT STOPS

HALL EFFECT = DIVERSION OF SOME CHARGE CARRIERS PERPENDICULAR TO DIRECTION OF CURRENT

THE HALL EFFECT—A TEST FOR SEMICONDUCTING CURRENTS

We tranquilized a salamander lightly, placed it on a plastic shelf between the poles of a strong electromagnet, and attached electrodes to measure the EEG. As we gradually increased the magnetic field strength, we saw no change—until delta waves appeared at 2,000 gauss. At 3,000 gauss, the entire EEG was composed of simple delta waves, and the animal was motionless and unresponsive to all stimuli. Moreover, as we decreased the strength of the magnetic field, normal EEG patterns returned suddenly, and the salamander regained consciousness *within seconds.* This was in sharp contrast to other forms of anesthesia.

With direct currents, the EEG continued to show delta waves for as long as a half hour after the current was turned off, and the animals remained groggy and unresponsive just as after chemical anesthesia.

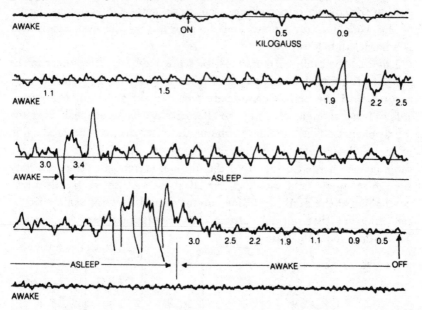

THE EEG OF A SALAMANDER ANESTHETIZED BY A MAGNETIC FIELD

It seemed to us that we'd discovered the best possible anesthetic, allowing prompt recovery with no side effects. We proposed getting a bigger electromagnet to try this method on larger animals and eventually humans, but we never even got a reply. Our data on direct currents in the nerves weren't quite acceptable, but reactions by living things to magnetic fields were absolutely out of the question in America at that time.

I was flabbergasted, therefore, to receive a phone call from one of the most prominent biologists in the Harvard-MIT orbit. He told me, "We're in the process of setting up an international conference on high-energy magnetic fields at MIT, and we've received a number of questions from respectable scientists in other countries asking why there is to be no session on biological effects of magnetic fields. This is a totally new idea to us, and we really don't believe there are such effects, but some of these fellows are persistent. We've searched the scientific literature and found your paper on the Hall effect. Since you seem to be the only person in this country doing any work along these lines, let me ask if you think there's anything to it."

I allowed as humbly as I could that there just might be something there, and told him about the latest experiments. There was a long pause filled with disdain. Then I added that Professor Bachman had been working with me. That changed the tone dramatically; this man also knew Charlie through work on the electron microscope during the war. He asked if I would organize the session and arrange for additional speakers.

There weren't too many investigators to choose from, and some of them were doing very slipshod work. I invited Frank Brown and selected a few others on the basis of published work. I'd just about finished when I got another call. A man with a thick German accent introduced himself as Dietrich Beischer.

"I have read your paper on the Hall effect," he began, "and I think we have much common interest." He explained that he was studying magnetic field bioeffects for the Navy and had done much work that wasn't published openly. At the time he was conducting a large experiment on human volunteers to check for effects from a null field, a complete absence of magnetism. When I wondered how he produced such a state, he invited me to have a look and perhaps offer suggestions. So off I went to Maryland.

Beischer was using the compass calibration building in the Naval Surface Weapons Center at Silver Spring. The building was huge. Electrical cables in all the walls, floor, and roof were "servo-connected" (directly cued) to the three axes of the earth's magnetic field, so that the field was canceled out in a sphere about 20 feet across in the center of the structure. Several men were living and being tested in this area. I was impressed by the resources at Beischer's command, and I had a good time, but I wondered what use any discoveries made there might ultimately be put to. My only contribution was to point out that the enclosure had been built before anyone knew about the earth field's low-frequency components, micropulsations ranging from less than 1 to about 25 cycles per second, that were far weaker than the planet's electromagnetic field as a whole. Consequently Beischer's subjects were still exposed to a very weak magnetic field pulsing at these frequencies, and I suspected that that component might be one of the most important for life, because all brain waves were in exactly the same range. Perhaps as a result of this factor, the null-field experiment was turning out to be inconclusive, but I asked Beischer to attend the MIT meeting and present some previous data suggesting that electromagnetic fields could affect embryonic development.

For my own presentation I decided against trying to compress all the work I'd done so far into a few minutes before a skeptical audience. Instead I offered some evidence that Charlie and I had gathered with the help of psychiatrist Howard Friedman, which showed a possible relationship between mental disturbances and solar magnetic storms. I'll discuss this study further in Chapter 14.

The MIT meeting went well. The field of bioelectromagnetism was still young, and the researchers in it didn't make many converts among the mainstream biologists. As usual, we found the physicists more inclined to listen. However, we drew inspiration from each other. I returned to the lab more determined than ever to elucidate the links that I knew existed between electromagnetic energy and life.

Charlie, Howard, and I decided to find out how the brain's DC potentials behaved in humans. The electrodes we'd been using on salamanders couldn't be scaled up for people, but within a week Charlie invented some that would give us equally precise readings from the human head. We immediately found that the back-to-front current varied with changes in consciousness just as in salamanders. It was strongest during heightened physical or mental activity, it declined during rest, and it reversed direction in both normal sleep and anesthesia. This knowledge led directly to the experiments, described in Chapter 13, that taught us much about how hypnosis and pain perception work.

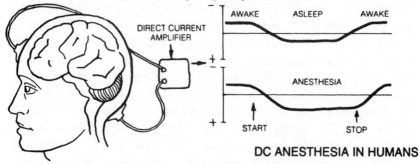

DC ANESTHESIA IN HUMANS

At this point I received an invitation from Meryl Rose to speak at *the* big event in the world of animal science, the International Congress of Zoology. This is not just a yearly convention; it is convened only when its directors agree that there has been enough progress to warrant a meeting. This session, in August 1963, was only the sixteenth since the first one had been set up in 1889. It was an honor to be there, and the conference itself was especially important, since it was one of the first times science formally addressed such ecological emergencies as pesticide pollution, the protection of vanishing species, overpopulation, and urban sprawl. The high point for me came when I gave my paper and saw in the audience Dr. Ralph Bowen, my college biology professor, a kind but exacting teacher who'd inspired me with his unique combination of scientific discipline and respect for life. Afterward, with characteristic caution, he said something like "That's not too bad, Becker. I'd like to see you keep going in this research."

When I assured him that, despite my M.D. degree, I was still committed to basic biology, he said, "I hope so, but remember, it won't be easy. To change things is never popular." His encouragement meant a lot to me, and I was happy to be able to show him that I'd amounted to something.

A lot had happened in the four work-filled years since I'd begun studying the current of injury. That first experiment had opened a door into a great hall with passageways leading off in all sorts of fascinating directions. This was really the life! Without leaving the laboratory I'd gone on a journey of exploration as exciting as trekking through the uncharted wilds of New Guinea. Our work on nerves and the brain was leading toward a whole new concept of life whose implications only gradually became apparent. Meanwhile, my colleagues and I were continuing to investigate the processes of healing, leading to insights and practical applications that more than justified my enthusiasm.

Six

The Ticklish Gene

Despite my fascination with fundamental questions about the nature of life, I was, after all, an orthopedic surgeon, and I was eager to find things that would help my patients. In addition, to convince the Artifact Man and all his brothers, Charlie and I were looking for some direct test of semiconduction in living tissues. The Hall effect and the freezing of frog nerves each demonstrated a characteristic of semiconduction but didn't confirm it in standard engineering terms. Unfortunately, all the direct tests then known worked only with crystals. You needed a material you could carve into blocks, something that didn't squish when you put an electrode on it. The only possibility was bone.

To many biologists and physicians, bones are pretty dull. They seem like a bunch of scarecrow sticks in which nothing much happens, plain props for a subtler architecture. Many of my patients were in sad shape because doctors had failed to realize that bone is a living tissue that has to be treated with respect. It's a common misconception that orthopedic surgery is like carpentry. All you have to do is put a recalcitrant fracture together with screws, plates, or nails; if the pieces are firmly fixed after surgery, you're done. Nothing could be further from the truth. No matter how firmly you hold them together, the pieces will come loose and the limb can't be used if the bone doesn't heal.

The Pillars of the Temple

The skeleton doesn't deserve this cavalier treatment. The development of bones by the first true fishes of the Devonian era nearly 400 million years ago was a remarkable achievement. It enabled animals to advance, in both senses of the word, quickly and efficiently. Since bone is inside the body, it can live and grow with the animal, instead of leaving it defenseless as an external skeleton does when cast off during a molt. It's also the most efficient system for attaching muscles and increasing the size of animals.

Bone is extraordinary in structure, too. It's stronger than cast iron in resisting compression but, if killed by X rays or by cutting off its blood supply (barely adequate to start with), it collapses into mush. The part that's actually alive, the bone cells, comprises less than 20 percent of the whole. The rest, the matrix, isn't just homogeneous concrete, either. It's composed of two dissimilar materials—collagen, a long-chain, fibrous protein that's the main structural material of the entire body, and apatite, a crystalline mineral that's mainly calcium phosphate. The electron microscope shows that the association between collagen and apatite is highly ordered, right down to the molecular level. The collagen fibers have raised transverse bands that divide them into regular segments. The apatite crystals, just the right size to fit snugly between these bands, are deposited like scales around the fibers.

This intricacy continues at higher levels of organization. The collagen fibers lie side by side in layer upon layer wound in opposed spirals (a double helix) around a central axis. The bone cells, or osteocytes, are embedded in these layers, which form units a few millimeters long, called osteones. The center of each osteone has a small canal in which runs a blood vessel and a nerve. The osteones in turn are organized so as to lie along the lines of maximum mechanical stress, producing a bone of the precise shape best able to withstand the forces applied to it.

Bone has an amazing capacity to grow, which it does in three different ways. In childhood each long bone of the limbs has one or two growth centers, called epiphyseal plates. Each is a body of cartilage whose leading edge grows continuously while its trailing edge transforms into bone. When the bone is the right length, the process stops, and the remaining cartilage forms the bony knob, or epiphysis, at the end of the bone. The "closure" of the epiphyses is an index of the body's maturation.

Bone cannot heal. That sounds like a conundrum but it's literally true. Fractures knit because new bone made from *other* tissues unites the fracture ends. Although we sometimes speak of bone growth as part of fracture healing, the old, preexisting bone doesn't have the capacity to grow. As mentioned in Chapter 1, there are two tissues that produce new bone at a fracture site. One is the periosteum, the bone's fibrous covering. It's the cells of the periosteum's innermost layer that have the

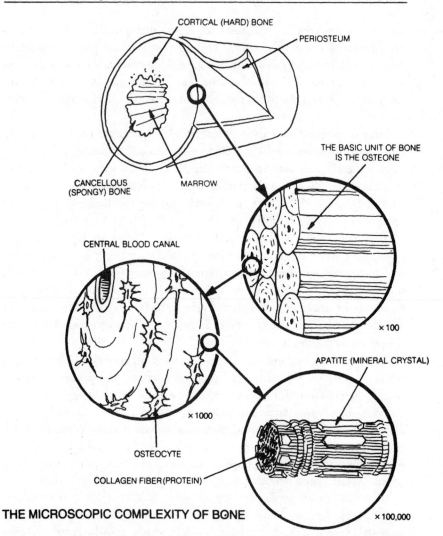

CORTICAL (HARD) BONE

PERIOSTEUM

THE BASIC UNIT OF BONE
IS THE OSTEONE

CANCELLOUS
(SPONGY) BONE

MARROW

CENTRAL BLOOD CANAL

× 100

APATITE (MINERAL CRYSTAL)

× 1000

OSTEOCYTE

COLLAGEN FIBER (PROTEIN)

× 100,000

THE MICROSCOPIC COMPLEXITY OF BONE

power of osteogenesis, or bone formation. After a fracture, these cells are somehow turned on. They begin to divide and some of the daughter cells turn into osteoblasts, cells that make the collagen fibers of bone. Apatite crystals then condense out of the blood serum onto the fibers.

The other tissue that forms new bone to heal a fracture is the marrow. Its cells dedifferentiate and form a blastema, filling the central part of the fracture. The blastema cells then turn into cartilage cells and later into more osteoblasts. This process is true regeneration, following the same sequence of cell changes as the regrowing salamander limb.

Whatever a physician does to repair a fracture, he or she must protect the periosteum and marrow cavity from harm. Unfortunately, too often

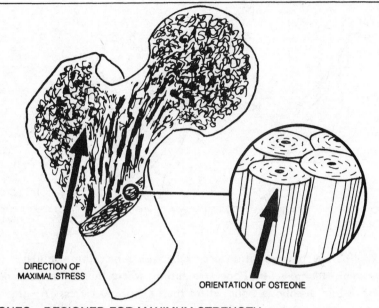

DIRECTION OF
MAXIMAL STRESS

ORIENTATION OF OSTEONE

BONES—DESIGNED FOR MAXIMUM STRENGTH

the application of plates, screws, and nails does just the opposite and, rather than helping nature, the treatment impedes healing.

From a researcher's point of view, the question here is: What activates the cells of the periosteum and marrow? In the case of the marrow, we can expect it to be the same factor that switches on the cells in a salamander's amputated leg.

There's a third process of growth that's unique to bone. It follows Wolff's law, which is named after the orthopedic surgeon J. Wolff, who discovered it at the end of the nineteenth century. Basically, Wolff's law states that a bone responds to stress by growing into whatever shape best meets the demands its owner makes of it. When a bone is bent, one side is compressed and the other is stretched. When it's bent consistently in one direction, extra bone grows to shore up the compressed side, and some is absorbed from the stretched side. It's as though a bridge could sense that most of its traffic was in one lane and could put up extra beams and cables on that side while dismantling them from the other. As a result, a tennis player or baseball pitcher has heavier and differently contoured bones in the racket arm or pitching arm than in the other one. This ability is greatest in youth, so in childhood fractures it's often best to put the bone ends together gently by manipulation without surgery, settling for a less than perfect fit. Sometimes the hardest part is convincing the parents that a modest bend will straighten itself out in a few months in accordance with Wolff's law.

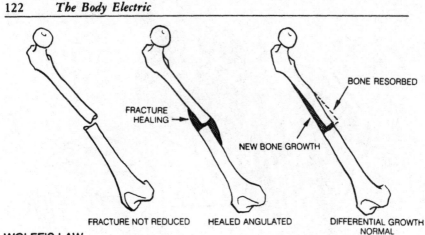

FRACTURE NOT REDUCED   HEALED ANGULATED   DIFFERENTIAL GROWTH
                                                      NORMAL

WOLFF'S LAW

Wolff's-law reorganization occurs because something stimulates the periosteum to grow new bone at a surface where there's compressional stress, while dissolving bone where there's tensional stress. Again, the question for researchers is: What turns on the periosteal cells?

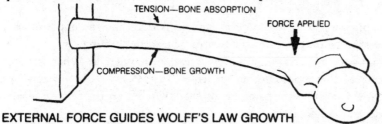

EXTERNAL FORCE GUIDES WOLFF'S LAW GROWTH

After I'd finished giving my paper on the salamander's current of injury at the orthopedic meeting early in 1961, several people came up to the stage to ask questions. Among them was Andy Bassett, a young orthopedic surgeon who was doing research at Columbia. In our conversation we came up with an angle for investigating Wolff's law—piezoelectricity. Simply put, this is the ability of some materials to transform mechanical stress into electrical energy. For example, if you bend a piezoelectric crystal hard enough to deform it slightly, there'll be a pulse of current through it. In effect, the squeeze pops electrons out of their places in the crystal lattice. They migrate toward the compression, so the charge on the inside curve of a bent crystal is negative. The potential quickly disappears if you sustain the stress, but when you release it, an equal and opposite positive pulse appears as the electrons rebound before settling back into place.

Since I'd shown that a stronger-than-normal negative current preceded regeneration, Bassett suggested that maybe bone was piezoelectric and the negative charge from bending stimulated the adaptive growth of Wolff's law. To find out, we tested both living and dead bones from a variety of animals, and found that bending produced an immediate potential, as expected. The compressed side became negative; the stretched side, positive.*

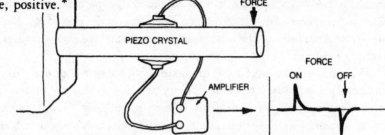

PIEZOELECTRICITY—MECHANICAL STRESS INTO ELECTRICITY

Furthermore, the reversed potentials that appeared when we released the stress weren't nearly as large as the first ones. This was just as it should be. If a negative voltage was the growth stimulus, there had to be some way to cancel out the positive rebound voltage; otherwise it would have negated the growth message. In electronic terms, there had to be a solid-state rectifier, or PN junction diode.

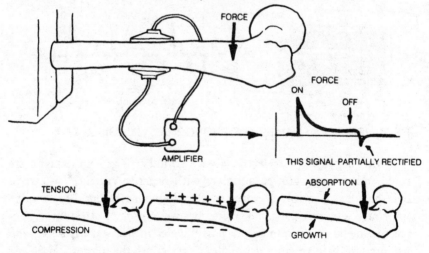

PIEZOELECTRICITY IN BONE

*After writing up this experiment, we found that it had been done before. Iwao Yasuda, a Japanese orthopedist, had shown that bone was piezoelectric back in 1954; he and Eiichi Fukada, a physicist, had confirmed the fact in 1957. We made note of their prior observations but published our paper anyway, since our techniques were different and ours was the first report in English.

Despite the initimidating names, this device is fairly simple. It's a filter that screens out either the positive (P) or the negative (N) part of a signal. As mentioned in Chapter 4, current can flow through a crystal lattice either as free electrons or as "holes" that can shift their positions much as the holes migrate when you move the marbles in a game of Chinese checkers. Since current can flow from a P-type to an N-type semiconductor but not the other way, a junction of the two will filter, or rectify, a current.

The phonograph would be impossible without this device. As the diamond or sapphire crystal of the stylus rides a record's groove, the groove's changing shape deforms it ever so slightly. The crystal, of course, transforms the stresses into a varying electrical signal, which is amplified until we can hear it. It would be an unintelligible hum, however, if we heard both the deformation pulse and the release pulse. Therefore we place a rectifier in the circuit. It passes current in one direction only, so the impulses don't cancel each other out. The signal is rectified, and when we feed it to a loudspeaker we hear music. Bassett and I felt sure we were seeing evidence of rectification in the fact that bone's release pulse was much smaller than the one from stress.

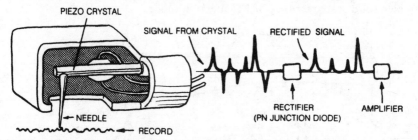

RECTIFIED SIGNALS MAKE THE PHONOGRAPH POSSIBLE

If the negative piezoelectric signal stimulated growth, maybe we could induce bone growth ourselves with negative current.* We tried out the idea on eighteen dogs. In the thighbone of one hind leg we implanted a battery pack. The electrodes were made of platinum, a non-reactive metal, to minimize any possible electrochemical irritation, and we inserted them through drill holes directly into the marrow cavity. As controls, some of the devices didn't have a battery. After three weeks we found that the active units had produced large amounts of new bone

---

*Again Dr. Yasuda and his colleagues had already done so, but their results seem to have been due to bone's ability to grow in response to irritation from the electrodes. They used alternating current, which is now known to have no direct growth-stimulating effect.

around the negative electrode but none around the positive. In the controls, there was no growth around either electrode.

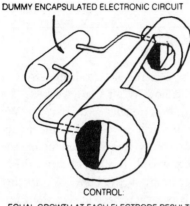

DUMMY ENCAPSULATED ELECTRONIC CIRCUIT   ACTIVE ENCAPSULATED ELECTRONIC CIRCUIT

CONTROL:
EQUAL GROWTH AT EACH ELECTRODE RESULTING
FROM STIMULUS OF ELECTRODE INSERTION

EXPERIMENT:
LESS GROWTH AT POSITIVE ELECTRODE,
MORE AT NEGATIVE ELECTRODE

ELECTRICAL STIMULATION OF BONE GROWTH

The results were exciting, but in retrospect I believe we made a serious error when we published this study. In our own minds and in print as well, we confused the negative potentials of the salamander's current of injury, the negative potentials that stimulated bone growth, and the negative potentials from the piezoelectric study. We proceeded as though they were equivalent, but they were not. The piezoelectric potentials were measured on the *outside* of the bone and appeared only when mechanical stress was applied. They were transitory, and most likely the periosteum was their target tissue. In the implant study, we used continuous direct current applied to the *inside* of the bone, the marrow cavity. What we were stimulating was the DC control system of regenerative fracture healing, not the piezoelectric control system of Wolff's law. We didn't clearly indicate the difference to the scientific reader, and this led to much confusion, some of which still persists twenty years later. As a result, many scientists think electricity stimulates bone growth because bone is piezoelectric. Most of these people don't realize that bone itself doesn't grow when a fracture heals. Moreover, everyone who has proceeded from our technique—and it's being used today to heal nonunions (see Chapter 8)—has done basically the same thing, continuously stimulating the bone marrow. No one has tried to stimulate the periosteum, as the pulsed piezoelectric signal does.

Our confusion also helped the scientific establishment accept the "trivial" electrical stimulation of bone by considering it something unique to bone. The relationship between our experiment and regeneration proper was lost.

The Inner Electronics of Bone

Charlie Bachman and I decided to investigate the electrical properties of bone in more detail and try to figure out how Wolff's law worked. We put together a hypothesis based on my experiments with Bassett. We postulated that the bone matrix was a biphasic (two-part) semiconductor. That is, either apatite or collagen was an N-type semiconductor; the other, a P type. Their connected surfaces would thus form a natural PN junction diode that would rectify any current in the bone. We further theorized that only one of the materials was piezoelectric. On the compressed side of a stressed bone, we expected the positive pulses to be filtered out, leaving a negative signal to stimulate periosteal cells to grow new bone.

We made several pairs of sample blocks, cut side by side from pieces of bone removed from patients for medical reasons. From one member of each pair we chemically removed the apatite. The other we treated with a compound that dissolved the collagen. The resulting pure collagen was yellowish and slightly rubbery, and the apatite pure white and brittle, but otherwise both blocks still looked like bone. Our first step was to test bone's component materials separately for semiconduction and piezoelectricity. Collagen turned out to be an N-type semiconductor and apatite a P type.

Then we tested our samples for piezoelectricity in the same way that Bassett and I had previously tested whole bone. We expected that apatite would be the only one to show an effect, since it was a crystal. However, collagen turned out to be a piezoelectric generator, while apatite was not. We now had the makings of a PN junction—two semiconductors, one an N type, the other a P type, joined together in a highly organized fashion.

Now came the crucial part of our hypothesis. We had to figure out a way to test for rectification at the PN junction. It was an important crossroads.

Here we ran up against what's known in the trade as a technical problem. To test for rectification we had to put one electrode on the collagen and one on the apatite *as they appeared in whole bone.* Unfortunately, the apatite crystals are each only 500 angstroms long. Now, the angstrom (named after the Swedish pioneer of spectroscopy Anders Jonas Ångström) was invented for measuring atoms and molecules, and it is not large. Five hundred of them are only one tenth as long as a single wave of green light. Even today's thinnest microelectrodes are 1 micron (10,000 angstroms) wide, and at the time the thinnest ones available to us were much larger. It would have been like trying to measure a grain of rice with a telephone pole.

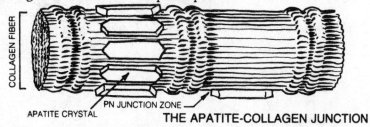

THE APATITE-COLLAGEN JUNCTION

We would have to do it in a sort of statistical way. Because of the way bone is built—millions of little scales glued onto larger fibers arranged in more or less lengthwise spirals along the osteones—I reasoned thus: If we put an electrode on the lengthwise cut, we'd be contacting mostly apatite, while an electrode on a face cut across the grain should connect to a greater proportion of collagen. If that method of electrode hookup worked and if we had a rectifier in our bones, then we'd be able to pass current through our samples only in one direction. That was exactly what happened. Our bone samples weren't as efficient as a commercial rectifier, but the amount of current we could put through them from a battery of constant voltage was much greater in one direction than in the other.

Current flowing "uphill," against the normal flow from P to N semiconductors, is called a reverse bias current, and we used it to look for photoelectric effects. Many semiconductors absorb energy from light, and any current flowing through the material gets a boost. We arranged our apparatus so only a small spot of light shone on the bone, because our silver electrodes were slightly sensitive to light and could produce a real artifact. With the voltage constant, the light produced an unmistakable increase in the current. Now, if bone really contained a rectifier, the photoelectric effect should be sensitive to the current's direction. The current in reverse bias should rise more with the same light intensity than the current in forward bias. The experiment was simple. We reversed the battery and turned on the light. The amperage rose higher than before. The rectifier was real.

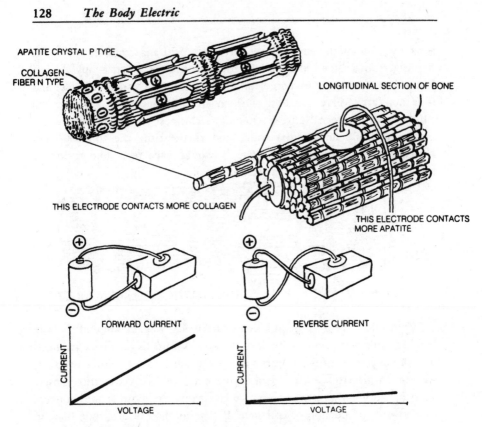

APATITE CRYSTAL P TYPE

COLLAGEN
FIBER N TYPE

LONGITUDINAL SECTION OF BONE

THIS ELECTRODE CONTACTS MORE COLLAGEN

THIS ELECTRODE CONTACTS
MORE APATITE

FORWARD CURRENT

REVERSE CURRENT

CURRENT

VOLTAGE

CURRENT

VOLTAGE

BONE IS A RECTIFIER—A PN JUNCTION DIODE

We could now follow the entire control system of Wolff's law. Mechanical stress on the bone produced a piezoelectric signal from the collagen. The signal was biphasic, switching polarity with each stress-and-release. The signal was rectified by the PN junction between apatite and collagen. This coherent signal did more than merely indicate that stress had occurred. Its strength told the cells how strong the stress was, and its polarity told them what direction it came from. Osteogenic cells where the potential was negative would be stimulated to grow more bone, while those in the positive area would close up shop and dismantle their matrix. If growth and resorption were considered as two aspects of one process, the electrical signal acted as an analog code to transfer information about stress to the cells and trigger the appropriate response.

Now we knew how stress was converted into an electrical signal. We had discovered a transducer, a device that converts other forces into electricity or vice versa. There was another transducer in the Wolff's-law system—the mechanism that transformed the electrical signal into appropriate cell responses. Our next experiment showed us something about how this one worked.

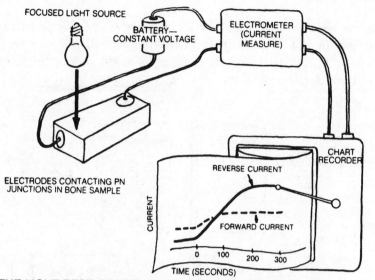

FOCUSED LIGHT SOURCE

BATTERY—
CONSTANT VOLTAGE

ELECTROMETER
(CURRENT
MEASURE)

CHART
RECORDER

ELECTRODES CONTACTING PN
JUNCTIONS IN BONE SAMPLE

REVERSE CURRENT

CURRENT

FORWARD CURRENT

0   100   200   300

TIME (SECONDS)

THE LIGHT TEST CONFIRMS BONE'S ELECTRICAL SYSTEM

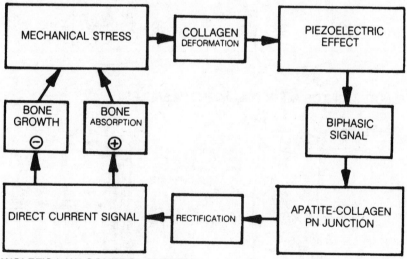

MECHANICAL STRESS

COLLAGEN
DEFORMATION

PIEZOELECTRIC
EFFECT

BONE
GROWTH
⊖

BONE
ABSORPTION
⊕

BIPHASIC
SIGNAL

DIRECT CURRENT SIGNAL

RECTIFICATION

APATITE-COLLAGEN
PN JUNCTION

WOLFF'S LAW CONTROL SYSTEM

Collagen fibers are formed from long sticks, like uncooked spaghetti, of a precursor molecule called tropocollagen. This compound, much used in biological research, is extracted from formed collagen—often from rat tails—and made into a solution. A slight change in the pH of the solution then precipitates collagen fibers. But the fibers thus formed are a jumbled, feltlike mass, nothing like the layered parallel strands of bone. However, when we passed a very weak direct current through the solution, the fibers formed in rows perpendicular to the lines of force around the negative electrode. This fit our new discoveries perfectly, because the lines of force on the negative (compressed) side of a bent bone would be in precisely the same alignment as the collagen fibers of the new bone that formed there.

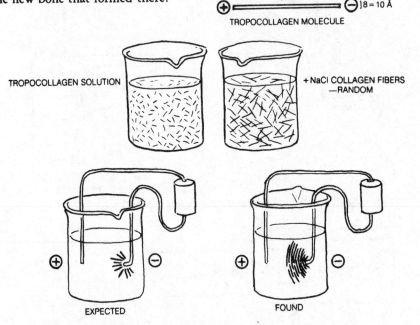

ELECTRIC FIELDS ALIGN COLLAGEN MOLECULES

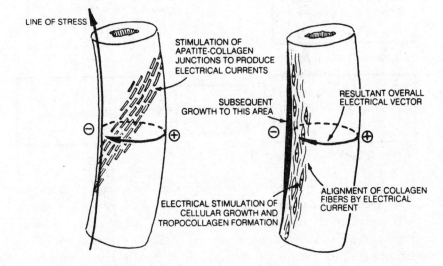

BONE'S ELECTRICAL SYSTEM GUIDES GROWTH

This was the first time that a complete circuit diagram of a growth process could be made. To us, this seemed an achievement of some note, but, although we got published, no one seemed to pay any attention. The scientific community wasn't ready for biological semiconduction, and the notion of diodes in living tissue seemed ridiculously farfetched to most of the people I told about it. For that reason, I never even bothered to try publishing one of our follow-up experiments. It was just too weird.

In the mid-1960s, solid-state devices were only beginning to hit the market, and one of the PN junction's most interesting properties hadn't yet been exploited. When you run a current through it in forward bias, some of its energy gets turned into light and emitted from the surface. In other words, electricity makes it glow. Nowadays various kinds of these PN junctions, called light-emitting diodes (LEDs), are everywhere as digital readouts in watches and calculators, but then they were laboratory curios.

We found that bone was an LED. Like many such materials, it required an outside source of light before an electric current would make it release its own light, and the light it emitted was at an infrared frequency invisible to us, but the effect was consistent and undeniable.

Even though we'd already proved our hypothesis, Charlie and I did a few more experiments on bone semiconduction, partly for additional confirmation and partly for the fun of it. It was known that some semiconductors fluoresced—that is, they absorbed ultraviolet light and emitted part of it at a lower frequency, as visible light. We checked, and whole bone fluoresced a bluish ivory, while collagen yielded an intense blue and apatite a dull brick-red. Here we found a puzzling discrepancy, however, which eventually led to a discovery that could benefit many people. When we combined the fluoresced light from collagen with that from apatite, we should have gotten the fluoresced light from whole bone. We didn't. That indicated there was some other material in the bone matrix, something we'd been washing out in the chemical separation.

Charlie and I were stumped on that line of research for a couple of years, until our attention was caught by a new development in solid-state technology called doping. Tiny amounts of certain minerals mixed into the semiconductor material could change its characteristics enormously. The making of semiconductors to order by selective doping would become a science in itself; to us it suggested trace elements in bone. We already knew that certain trace metals—such as copper, lead, silver, and beryllium—bonded readily to bone. Beryllium miners had a high rate of osteogenic sarcoma—bone cancer—because beryllium somehow removed normal controls from the osteocytes' growth potential. Radioactive strontium 90 worked its harm by bonding to bone, then bombarding the cells with ionizing radiation. Perhaps some trace element normally found in bone changed its electrical properties by doping it.

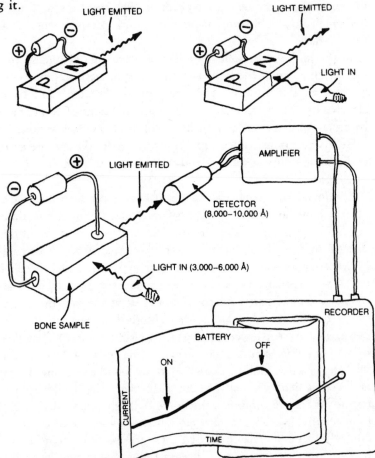

PHOTOELECTRICITY: ELECTRICITY GOES IN; LIGHT COMES OUT

To find out, Charlie and I used a very complex device called an electron paramagnetic resonance (EPR) spectrometer on our bone samples. There's no easy way to explain just how this instrument works, but basically it measures the number of free electrons in a material by sensing a resonance produced in the electrons' vibrations by an applied magnetic field. We used it to measure the free electrons in collagen and apatite, and we found the same kind of discrepancy as in our fluorescence experiment: When we added together the free electrons of collagen and apatite, we fell short of the number we found in whole bone. That made us certain that we were washing out some trace mineral.

We decided to work backward. We prepared a solution containing small amounts of a wide variety of metals. Then we soaked our collagen and apatite cubes in this broth to see what they'd pick up.

We knew we were on the way to solving this mystery when we examined the results. Only a few of the metals had bonded to the bone materials: beryllium, copper, iron, zinc, lead, and silver. The diameters of all the absorbed atoms were exact fractions of one another. The results showed that the bonding sites were little recesses into which would fit one atom of silver or lead, two of iron or copper or zinc, or six of beryllium.

Only one of these metals gave us an electron resonance of its own, indicating that it had a large number of free electrons that could affect the electrical nature of bone. That metal was copper. We made a batch of broth containing only copper. We expected that copper's EPR signal would change to one value as it bonded with collagen, and to another as it bonded with apatite. Since the molecular structure of each was quite different, we figured that each would bind copper in a different way.

We could hardly believe the results. Bonding had indeed changed copper's resonance, but the change was the same in both materials. By analyzing it we deduced that each atom of copper fit into a little pit, surrounded by a particular pattern of electric charges, on the surface of apatite crystals and collagen fibers. Because the pattern of charges was the same in both materials, we knew that the bonding sites were the same on both surfaces and that they lined up to form one elongated cavity connecting the crystal and fiber. In other words, the two bonding sites matched, forming an enclosed space into which two atoms of copper nestled. The electrical forces of this copper bond held the crystals and fibers together much as wooden pegs fastened the pieces of antique furniture to each other. Furthermore, the electrical nature of peg and hole suggested that we had found, on the atomic level, the exact location of the PN junction.

This discovery may have some medical importance. The question of how the innermost apatite crystals fasten onto collagen had eluded orthopedists until then, and the finding may have opened a way to understand osteoporosis, a condition in which the apatite crystals fall off and the bone degenerates. The process is often called decalcification, although more than calcium is lost. It's a common feature of aging. I surmise that osteoporosis comes about when copper is somehow removed from the bones. This might occur not only through chemical/metabolic processes, but by a change in the electromagnetic binding forces, allowing the pegs to "fall out." It's possible that this could result from a change in the overall electrical fields throughout the body or from a change in those surrounding the body in the environment.

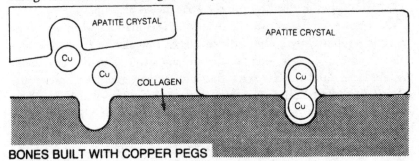

BONES BUILT WITH COPPER PEGS

Osteoporosis has been a major worry of the American and Russian space programs. As flights got longer, doctors found that more and more apatite was lost from the bones, until decalcification reached 8 percent in the early Soviet Salyut space-station tests. Serious problems were known to occur only when apatite loss reached about 20 percent, but the trend was alarming, especially since depletion of the calcium reservoir might affect the nervous system and muscular efficiency before the 20-percent level was reached. Although the immediate cause was their inability to get a malfunctioning air valve closed before all the cabin air escaped, weakness from muscle tone loss might have contributed to the death of the three cosmonauts who succumbed while returning from their twenty-four-day flight aboard Soyuz 11 at the end of June 1971.

Space osteoporosis may result from unnatural currents induced in bone by a spacecraft's rapid motion through the earth's magnetic field, with a polarity reversal every half orbit, or it may be a direct effect of the field reversal. This abnormality, which may change the activity of bone cells directly, would be superimposed on abnormal responses of bone's natural electrical system, which is almost certainly affected by weightlessness. The unfamiliar external field reversals could also weaken the copper pegs, at the same time that the bones are in a constant state of "rebound" from their earthly weight-induced potentials, producing a signal that says, "No weight, no bones needed." We know that the more even distribution of blood caused by weightlessness registers in the heart as an excess; as a result, fluid and ions, including calcium, are withdrawn from the blood. However, the effect probably isn't due to weightlessness alone, for the Skylab astronauts did rigorous exercise, which would have supplied plentiful stresses to their bones. They worked out so hard that their muscles grew, but decalcification still reached 6.8 percent on the twelve-week mission.

The Soviets at first claimed to have solved the problem before or during the Soyuz 26 mission of 1977–78, in which two cosmonauts orbited in the Salyut 6 space lab for over three months. Subsequent Soviet spacepersons, who have remained weightless as long as 211 days, reportedly have showed no ill effects from osteoporosis, and chief Soviet space doctor Oleg Gazenko said it simply leveled off after three months. However, this claim was later officially withdrawn, and I have a hunch that the Soviets are working on a way to prevent the condition by simulating earth-surface fields inside their space stations, a method that perhaps hasn't yet worked as well as they'd hoped. Andy Bassett has suggested giving our astronauts strap-on electromagnetic coils designed to approximate their limb bones' normal gravity stress signals, but so far NASA has shown no interest.

Unfortunately for earthbound victims of osteoporosis, the copper peg discovery still hasn't been followed up, even though I published it over fifteen years ago. Charlie and I wanted to continue in that direction, but we knew that we couldn't sustain more than one major research effort at a time. We decided that regenerative growth control was our primary target, so we reluctantly dropped osteoporosis. Fortified by our new knowledge that electricity controlled growth in bone, we returned instead to the nerves, taking a closer look at how their currents stimulated regrowth.

A Surprise in the Blood

I felt as though the temple curtain had been drawn aside without warning and I, a goggle-eyed stranger somehow mistaken for an initiate, had been ushered into the sanctuary to witness the mystery of mysteries. I saw a phantasmagoria, a living tapestry of forms jeweled in minute detail. They danced together like guests at a rowdy wedding. They changed their shapes. Within themselves they juggled geometrical shards like the fragments in a kaleidoscope. They sent forth extensions of themselves like the flares of suns. Yet all their activity was obviously interrelated; each being's actions were in step with its neighbors'. They were like bees swarming: They obviously recognized each other and were communicating avidly, but it was impossible to know what they were saying. They enacted a pageant whose beauty awed me.

As the lights came back on, the auditorium seemed dull and unreal. I'd been watching various kinds of ordinary cells going about their daily business, as seen through a microscope and recorded by the latest time-lapse movie techniques. The filmmaker frankly admitted that neither he nor anyone else knew just what the cells were doing, or how and why they were doing it. We biologists, especially during our formative years in school, spent most of our time dissecting dead animals and studying preparations of dead cells stained to make their structures more easily visible—"painted tombstones," as someone once called them. Of course, we all knew that life was more a process than a structure, but we tended to forget this, because a structure was so much easier to study. This film reminded me how far our static concepts still were from the actual business of living. As I thought how any one of those scintillating cells potentially could become a whole speckled frog or a person, I grew surer than ever that my work so far had disclosed only a few aspects of a process-control system as varied and widespread as life itself, of which we'd been ignorant until then.

The film was shown at a workshop on fracture healing sponsored by the National Academy of Sciences in 1965. It was one of a series of meetings organized for the heads of clinical departments to educate them as to the most promising directions for research. A dynamic organizer named Jim Wray had recently become chairman of the Upstate Medical Center's department of orthopedic surgery, but Jim's superb skills were political rather than scientific. Since I was an active researcher and had just been promoted to associate professor, Jim asked me to go in his place. I tried to get out of it, because I knew my electrical bones would get a frosty reception from the big shots if I opened my mouth, but Jim prevailed. The meeting was mostly what I'd expected, but there were three bright spots. One was that microcinematic vision. Another was the chance to get acquainted with the other delegate from my department, a sharp young orthopedic surgeon named Dave Murray. The third was the presence of Dr. John J. Pritchard.

A renowned British anatomist who'd added much to our knowledge of fracture healing, Dr. Pritchard was the meeting's keynote speaker—the beneficent father-figure who was to evaluate all the papers and summarize everything at the end. Dave and I almost skipped his talk. We hadn't been impressed with the presentations, and we figured there had been so few new ideas that Pritchard would have nothing to say. However, our bus to the Washington airport didn't leave until after Pritchard's luncheon address, so we stayed. With a tact that seemed peculiarly English, he reached the same assessment we had, but phrased it so as to offend no one. He stressed that fracture healing should be considered a vestige of regeneration. Most past work on fractures had described *what* happened when a bone knit, as opposed to the how and why. As Pritchard pointed out, "Not a great deal of thought has been given to the factors that initiate, guide, and control the various processes of bone repair." Just as in regeneration research, this was the most important problem about fractures, he concluded.

Dave and I had to wait several hours before our flight back to Syracuse. We sat in the airport lounge and talked excitedly about broken bones. Dave agreed that, since I'd found electrical currents in salamander limb regeneration, it was at least plausible that similar factors controlled the mending of fractures. Having just deciphered the control system for stress adaptation (Wolff's-law growth) in bone, I felt prepared to get back to the more complex problems of regeneration via its remnant in bone healing. Dave and I decided to collaborate, and we planned our experiment on the plane. We would break the same bone in a standardized way in each of a series of experimental animals. I would study the electrical forces in and around the fractures as they healed. We would kill a few of the animals at each stage of healing, and Dave, an expert histologist (cell specialist), would make microscope slides of the healing tissues and study the cellular changes. Along the way we would fit our findings together to see whether electricity was guiding the cells.

Our first task was to choose the experimental animals. We wanted to use dogs or rabbits, since ultimately we were trying to understand human bones and wanted to work with animals as closely related to us as possible. But we would need scores of them to study each phase of healing adequately, and we had neither the funds nor the facilities to house so many large mammals. We thought of rats, but their longest bones were too short to study clearly and were curved as well. We were looking for nice, long, straight bones, in which we could produce uniform breaks.

We settled on bullfrogs. They were cheap to buy and care for; we could even collect some ourselves from nearby ponds. I already had a lot of experience in working with them. Best of all, the adult frog's lower leg had one long bone—the tibia and fibula found in most vertebrates had merged into a tibiofibularis. It was about two inches long with a fine, straight shaft.

Our misgivings about the evolutionary distance between frogs and humans were allayed when we went to the library to read up on what was then known about fracture healing in frogs. Dr. Pritchard himself, along with two of his students, J. Bowden and A. J. Ruzicka, had determined that frogs mended their bones the same way people did. Our question was: What stimulated the periosteal and marrow cells to change into new bone-forming cells?

We began by anesthetizing the animals and resolutely breaking all those little green legs by hand, bending them only to a certain angle so as not to rupture the periosteum around the fracture. I found I had to put little plaster casts on them—not because the frogs seemed in great pain but because their movements kept shifting the broken bones and making systematic observations impossible. They would have healed anyway; in our first sixty frogs we found two that had broken their legs in the wild and mended them, but I'm sure ours were the first that ever had casts.

The electrical changes were complex but were almost the same in every fracture. There were two distinct patterns, one on the periosteum and one on the bone. Before fracture the ankle end of both the bone and the periosteum had a small negative potential of less than 1 millivolt as compared to the knee end. At the moment of fracture, the negative potential on the intact periosteum over the break shot up to 6 or 7 millivolts, while areas of positive charge formed above and below the break. After a week, the periosteum's normal progression of negative charge toward the ankle was restored. When a fracture ruptured the periosteum, its negative potential went even higher than 7 millivolts, but amputation of the dangling lower leg immediately reversed the polarity, producing a positive current of injury from the stump, as in the frogs of my first regeneration experiment. The bone itself underwent a short-term electrical change opposite to that in the periosteum. A small positive charge appeared on each of the broken ends during the first hours, then fell to near zero after three hours.

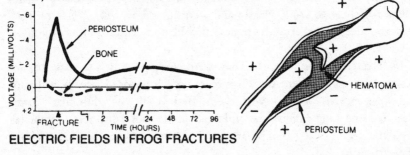

ELECTRIC FIELDS IN FROG FRACTURES

The electricity had two different sources. When I cut the leg nerves, the periosteum readings dropped dramatically, indicating that these potentials were coming from currents in the nerves to the periosteum and the surrounding wound area. Measurements on the bone, which has almost no nerves, were unaffected. Many piezoelectric materials emit a continuous current for several hours after their charge-producing structure has been left in a state of unresolved stress by fracture; I surmised that this was true of bone and soon found that another research team had recently proven it. Two separate currents, then, one from the nerves and one from the bone matrix, were producing potentials of opposite polarity, which acted like the electrodes of a battery. These living electrodes were creating a complex field whose exact shape and strength reflected the position of the bone pieces. The limb was, in effect, taking its own X ray.

While I was busy with my probes and meters, Dave was taking samples of the bones and blood clots and preparing them for the microscope. We killed a few of the frogs every fifteen minutes during the first two hours, then every day for two weeks, every other day for the third week, and weekly for the last three weeks. Preparing the slides took several days.

In the normal sequence of bone healing in frogs, a blood clot forms after about two hours and develops into a blastema during the first week. It turns into the rubbery, fibrous callus during the second and third weeks, and ossifies in three to six weeks. In this last period, islands of bone first emerge near the broken ends. Next, bony bridges appear, connecting the islands. Then the whole area is gradually filled in and organized with the proper marrow space and blood canals to join the segments of old bone.

Dave began his work with specimens taken nearly a week after the fractures, when we expected to see the first signs of the callus forming. "This is mighty damn funny," he said as he walked in with the first box of slides. "I can't see any mitoses in the periosteum. There's no evidence that the cells there are multiplying or migrating."

We agreed that we must have done something wrong. Pritchard's work had been quite conclusive on this point. He'd even published photographs of the periosteal cells dividing and moving into the gap. We thought maybe we were looking at specimens from the wrong time period, but we could see with our own eyes that the callus *was* starting to form. Dave went back to study specimens from the first few days, even though we didn't expect to see much then except clotting blood. Soon he called me from his lab and asked, "What would you say if I told you that the red corpuscles change and become the new bone-forming cells?"

I groaned. "Nonsense. That can't be right." But it *was* right. We went over the whole slide series together. Beginning in the second hour, the erythrocytes (red cells) began to change.

All vertebrates except mammals have nuclei in their red blood cells. In mammals, these cells go through an extra stage of development in which the nucleus is discarded. The resulting cells are smaller, can flow through smaller capillaries, can be packed with more hemoglobin, and thus can carry oxygen and carbon dioxide more efficiently. Nucleated erythrocytes are considered more primitive, but even in these the nucleus is pycnotic—shriveled up and inactive. The DNA in pycnotic nuclei is dormant, and such cells have almost no metabolic activity; that is, they burn no glucose for energy and synthesize no proteins. If you had to choose a likely candidate for dedifferentiation and increased activity, this would be the worst possible choice.

In our series of slides the red cells went through all their developmental stages in reverse. First they lost their characteristic flattened, elliptical shape and became round. Their membranes acquired a scalloped outline. By the third day the cells had become ameboid and moved by means of pseudopods. Concurrently, their nuclei swelled up and, judging by changes in their reactions to staining and light, the DNA became reactivated. We began using an electron microscope to get a clearer view of these changes. At the end of the first week, the former erythrocytes had acquired a full complement of mitochondria and also ribosomes (the organelles where proteins are assembled), and they'd gotten rid of all of their hemoglobin. By the third week they'd turned into cartilage-forming cells, which soon developed further into bone-forming cells.

I wasn't happy with this turn of events. How could we reconcile what we saw with the well-documented findings of Pritchard, Bowden, and Ruzicka? I'd expected evidence for the semiconducting electrical system I'd been investigating, a concept that was already strange enough to keep me out of the scientific mainstream. I would have been happy if the electrical measurements had fit in with straightforward changes in the periosteal cells. The difference between them and erythrocytes was crucial. Periosteal cells were closely related precursors of bone cells; blood cells couldn't have been further removed. They couldn't possibly have built bone without extensive job retraining on the genetic level. These bullfrogs were bringing us up hard against a wall of dogma by showing us metaplasia—dedifferentiation followed by redifferentiation into a totally unrelated cell type. The process took place in some of the *most* specialized cells in the frog's entire body, and it looked as though the electric field set the changes in motion while at its strongest, about an hour after the fracture.

Our next move was a respectful letter to Dr. Pritchard asking if there was any way he could make sense of the contradictory observations. He replied in the negative but sent our inquiry to Dr. Bowden, who had done the actual work on frogs as his doctoral thesis. Bowden had a possible explanation. He'd done the experimental work under a time limit, and to finish before the deadline he'd kept his frogs at high temperatures—only a few degrees short of killing them, in fact—in order to speed up their metabolism.

Bowden also mentioned that two researchers cited in his bibliography had seen fracture healing in frogs much the same way we had. In the 1920s, a German named H. Wurmbach, also working on his doctorate, noted some strange cellular transformations in the blood clot and worried over his inability to explain them. However, Wurmbach also found mitoses in the periosteum and ascribed healing to the latter process, since it didn't involve dedifferentiation. A decade later, another German scientist, A. Ide-Rozas, saw the same changes in the blood cells, but he was more daring. He proposed that this transformation was the major force behind fracture healing in frogs and further suggested that regenerating salamanders formed their limb blastemas from nucleated red blood cells. Other experiments seemed to contradict Ide-Rozas' idea about limb blastemas, so his work was discredited and ignored, but Bowden wished us better luck.

Bowden's letter gave us a framework for understanding our results. We already knew that mammals did *not* heal bones by dedifferentiation of their red corpuscles, because their red cells had no nuclei and thus no mechanism for change. Mammals also had a thicker periosteum than other vertebrates, so we reasoned that periosteal cell division played a larger healing role in mammals. Frogs, it seemed, had both methods available but activated the periosteal cells only at high temperatures.

Do-It-Yourself Dedifferentiation

Now we were sure that our results were real. We repeated the same fracture studies, but this time we also observed the cells while they were alive. We took tissue samples from the fractures and made time-lapse film sequences using techniques like those in the movie that had impressed me so much at the NAS workshop. We confirmed that the changes began in the first few hours, just after the electrical forces reached their peak.

Now we decided to try a crucial test. If the electricity really triggered healing, we should be able to reproduce the same field artificially and start the same changes in normal blood cells outside the frog. If this didn't work, then I probably had spent the last seven years "collecting stamps"—accumulating facts that were interesting but, in the end, trivial.

I calculated the amount of current that would produce the fields I'd found. I came up with an incredibly small amount, somewhere between a trillionth and a billionth of an ampere (a picoamp and a nanoamp, respectively). Again I thought there must be some mistake. I didn't see how such a tiny current could produce the dramatic effects we'd seen, so, figuring that even if my numbers were right, more juice would simply hasten the process, I decided to start with 50 microamps, a current level that would be just shy of producing a little electrolysis—the breakdown of water into hydrogen and oxygen.

I designed plastic and glass chambers of various shapes, fitted with electrodes of several types. In these chambers we would place healthy red blood cells in saline solution and observe them by microscope while the current was on.

I set up the experiment in a lab across the street from the medical center, where there was available one of the inverted microscopes we would need to observe the cells through the bottoms of the chambers, where most of them would settle. I put a young technician named Frederick Brown in charge of the long grind of watching the cells hour after hour at different current levels and field shapes in the various chambers. We began in the summer of 1966. Fred was to enter medical school that fall, and I figured two months would be more than enough time. He was to run one test batch of frog blood each day and report to me the next morning as to what he'd found.

It didn't start well. Nothing had happened after six hours of current. We couldn't increase the amperage without electrocuting the cells, so we ran it longer. Still nothing happened. In fact, the cells started dying when we left them in the chambers overnight. We decided to lower the current, but I still didn't believe in the absurdly low values I'd calculated, so I told Fred to drop the amperage only a little bit day by day. He and I stared at a lot of blood cells over those two months, all stubbornly refusing to do anything. Finally, two days before Fred had to leave, we'd gotten the current down as far as our first apparatus could go, and well within the range I'd calculated—about half a billionth of an ampere. At eleven that morning he called me excitedly and I rushed across the street.

With the room darkened and the microscope light on, we saw the same cell changes as in the blood clot, first at the negative electrode, then at the positive electrode, and finally spreading across the rest of the chamber. In four hours all the blood cells in the chamber had reactivated their nuclei, lost their hemoglobin, and become completely unspecialized in form.

NORMAL FROG RED BLOOD CELLS

CURRENT GENERATOR

ELECTRODE

MICROSCOPE

NORMAL

RADIAL CONDENSATION OF CYTOPLASM

NUCLEAR CONDENSATION

RETRACTILE NUCLEUS

HEMOGLOBIN LOST

CYTOPLASM VACUOLIZATION

SPHERICAL

AMOEBOID

PRIMITIVE

THE FIRST ARTIFICIAL DEDIFFERENTIATION

We repeated the experiment many times, working out the upper and lower limits of the effective current. The best "window" was somewhere between 200 and 700 picoamps. I say "somewhere" because the susceptibility of the cells varied, depending on their age, the hormonal state of the frog, and possibly other factors.* This was an infinitesimal tickle of

* Rather than continually renewing a small part of their red-cell stock, frogs generate a whole year's supply in late winter as they emerge from hibernation. Thus, all their erythrocytes age uniformly as the year progresses. The cells become less sensitive to electricity as they get older, and that may be why frogs, even when warm and not hibernating, heal fractures more slowly in winter. The red blood cells dedifferentiate most readily in the spring, and the female's become even more sensitive than the male's when she ovulates early in that season. At that time her red corpuscles will despecialize in response to less than one picoamp. In fact, we saw red cells from ovulating females dedifferentiate completely in chambers to which we supplied no current whatever. Apparently an unmeasurably small current created by the charge difference between the plastic chamber and glass cover slip was enough.

electricity, far less than anything a human could feel even on the most sensitive tissue, such as the tongue, but it was enough to goose the cell into unlocking all its genes for potential use.

The effect depended on having the proper cells as well as the proper current—white blood cells, skin cells, and other types didn't work. Only erythrocytes seemed to serve as target cells in frogs. We found the same response in the blood cells of goldfish, salamanders, snakes, and turtles. The only variation was that the fish cells despecialized faster and the reptilian cells more slowly than frog blood cells. In all erythrocytes the shift in the transparency and staining characteristics of the nucleus was a point of no return. These changes seemed to indicate reactivation of the DNA, for afterward the rest of the process continued even if we switched off the current.

This was a breakthrough. We'd learned something hitherto unsuspected about fracture healing in frogs, and it was almost certain to benefit human patients a few years down the road. Because we'd used frogs instead of mammals, we'd also stumbled upon the best proof yet for dedifferentiation—a do-it-yourself method. If we'd studied fracture healing in mammals, we almost certainly would not have made the discovery, for periosteal cells don't dedifferentiate and marrow cells are hard to experiment with. Instead, we even had movies of dedifferentiation happening and electron photomicrographs of all its stages, including brand-new ribosomes being made in the nucleus and deployed into the surrounding cytoplasm. Moreover, all the steps in dedifferentiation, including the activities in the nucleus and the assembly of ribosomes and mitochondria, exactly paralleled the changes found by the most recent research on salamander limb blastemas. We'd found the electrical common denominator that started the first phase—the blastema—in all regeneration.

The Genetic Key

Soon after we'd finished this experiment, I was invited to a meeting on electromagnetism in biology at the New York Academy of Sciences. This was basically a one-man show. Kenneth MacLean, a prominent surgeon and highly placed member of the academy, had been using magnetic fields on his patients for years and was convinced that they helped. Independently wealthy, he'd set up a lab in his office, with a large electromagnet. The meeting was a testament to his persistence rather than any widespread belief within the academy that he was right. So in February 1967 I presented our recent work. I played down the full role of electricity in fracture healing, with its overtones of vitalism, and concentrated on our method for inducing dedifferentiation *in vitro*. That was enough to call forth many attacks from the audience, most of which were variations on "I just don't believe it." Some said we were just electrocuting the cells, despite the fact that they survived for ten days in culture.

One audience member responded with some thoughtful and constructive criticism, however. He accepted the fact that we'd seen what we'd seen. Nevertheless, there were, he said, other barely possible explanations. In particular, he stated, we hadn't gone far enough in proving that the cells weren't slowly degenerating from some small but harmful change caused by the current. While the cells in our chambers looked like those we'd photographed in the fracture clot, our idea that these cells were electrically dedifferentiated healing cells was so at variance with current views that we must have more direct proof. For such a radical departure, seeing wasn't quite believing.

Stimulated by this one honest reaction, Dave and I returned to Syracuse and planned how we could use the latest knowledge about DNA to test our evidence further. A few years before, James Watson and Francis Crick had proposed what became known as the central dogma of genetics. In simplified form, it stated that the active DNA in each specialized cell imprinted its own specific patterns onto transfer RNA, which relayed them to messenger RNA. This second RNA molecule moved outside the nucleus to the ribosomes, where it translated the genetic instructions into the particular proteins that made the cell what it was.

We reasoned that, since a dedifferentiating cell was not going to divide immediately, it wouldn't duplicate its genes. Therefore there should be no increase in the amount of DNA it contained. However, since the cell changed its type by manufacturing a whole new set of proteins, the amount of RNA—the protein blueprints—should increase dramatically.

Using radioactive labeling and fluorescent staining techniques, we found there was indeed no new DNA but dramatic increases in RNA. Another test showed that our despecialized cells contained not only different proteins but also twice as many as their precursor red blood cells.

The most conclusive experiment was one suggested by Dan Harrington, a student who'd taken Fred Brown's place and who later went on to a Ph.D. in anatomy. Dan proposed that we use certain well-known metabolic inhibitors that disrupt the DNA-RNA-protein system, to see if we could prevent dedifferentiation. One such inhibitor, an antibiotic called puromycin, blocks the transfer of information from messenger RNA to the ribosomes and thus prevents proteins from being built. Dan proposed that we set up our plastic chambers in pairs. In one member of each pair we would place blood cells in saline solution and pass current through them as before. The other chamber would contain cells from the same frog and be connected to the same generator, but the solution would contain puromycin. The setup would thus be precisely controlled. If the current was actually unlocking new genes, the puromycin should stop dedifferentiation by intercepting the DNA's protein-making instructions. If the current was merely making the cells degenerate, however, the puromycin should have no effect and the transformations should continue.

Our conclusion held up: The cells in the puromycin solution didn't change. Next Dan suggested that we bathe these cells with several changes of water to wash out the puromycin. They promptly dedifferentiated with no current flowing! Apparently the current caused the genetic change in spite of the antibiotic, and the genes stayed unlocked, so the system still worked perfectly as soon as the puromycin blockade was lifted.

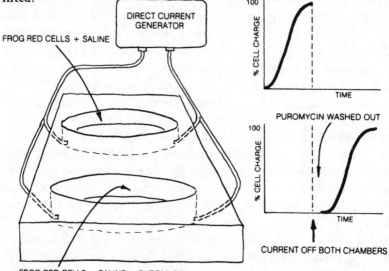

THE RNA BLOCKADE CONFIRMS ELECTRICAL DEDIFFERENTIATION

By early 1970 we had solid proof for nearly every detail of the control system for fracture healing in frogs, and by extension probably in mammals as well. Like all other injuries, a fracture produced a current of injury, in this case derived from the nerves in and around the periosteum.* At the same time, the bone generated its own current piezoelectrically due to residual stress in the mangled apatite-collagen matrix. These signals combined to stimulate the cells that formed new bone.

Except for the identity of the target cells, bone repair seemed to be basically the same in all vertebrates, proceeding through the stages of blood clot, blastema, callus, and ossification. In fish, amphibians, reptiles, and birds, the red cells in the clot dedifferentiated in response to the electric field, especially the positive potentials at the broken ends of the bone. They then redifferentiated as cartilage cells and continued on to become bone cells.

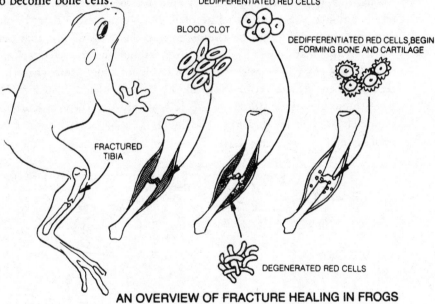

DEDIFFERENTIATED RED CELLS

BLOOD CLOT

DEDIFFERENTIATED RED CELLS, BEGIN FORMING BONE AND CARTILAGE

FRACTURED TIBIA

DEGENERATED RED CELLS

AN OVERVIEW OF FRACTURE HEALING IN FROGS

In some animals the periosteal cells responded to the current of injury by migrating into the gap and specializing a bit further into bone cells. This process seemed to be available to amphibians only at high temperatures, but it was the dominant method in mammals, whose thick periosteum made up for the lack of nuclei in their red corpuscles. By this time we'd become pretty sure that the marrow component of bone healing in humans involved the dedifferentiation of at least the *immature*

---

*The lack of the periosteal (nerve-derived) current may explain the uncontrolled, deformed growth that often follows fractures in the limbs of paraplegics and lepers. Their bones still generate a positive potential in the gap, but because of nerve damage it isn't balanced by the negative periosteal potential that normally surrounds the break.

erythrocytes, which still contained a nucleus, and possibly other cell types.

The electrical forces turned the key that unlocked the repressed genes. The exact nature of that key was the one part still missing from the process. The current couldn't act directly on the nucleus, which was insulated by the cell's membrane and cytoplasm. We knew that the current's primary effect had to be on the membrane. The cell membrane itself was known to be charged. Its charge probably occurred as a specific pattern of charged molecules, different for each type of cell. We postulated that the membrane released derepressors—molecules that migrated inward into the nucleus, where they switched on the genes. Based on recent findings about the structure of RNA, we suggested that the derepressor molecules might be a stable form of messenger RNA that persisted in the mature red cell even after its nucleus shriveled up and turned itself off. RNA molecules can be stable for a long time when they are folded, the strands secured together by electron bonds. If such folded RNA molecules were stored in the cell membrane, the tiny currents could release their bonds and unfold them. This hypothesis has not yet been tested.

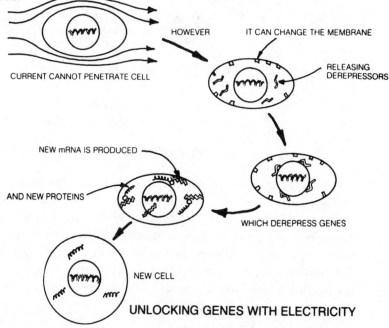

UNLOCKING GENES WITH ELECTRICITY

Fracture healing was ended by a straightforward negative-feedback system. As the gap was filled in with new matrix, the bone gradually redistributed its material to balance the stresses on it from the action of the surrounding muscles during cautious use, in accordance with Wolff's law. Repair of surrounding tissues lessened and then stopped the periosteal injury current. As a result the electrical field returned to normal, shutting off the cellular activities of healing.

When we'd finished this series of experiments, I was sure this was the most important piece of work I would ever do, and I was determined to get it published as a major article, not just a short note. Luck was with me. On several speaking engagements during the previous two years I'd been able to talk at length with Dr. Urist. He was enthusiastic about our findings, and since he was the editor of one of the major orthopedic journals—*Clinical Orthopedics and Related Research*—I submitted our report there. The editorial board published it uncut, and I was pleased to use Dr. Pritchard's statement at the 1965 bone-healing workshop as an epigraph. To me it's still the most satisfying of my publications.

This was the first time the control system for a healing process had been worked out in such detail. Except for the less conclusive account Dave and I had presented to the New York Academy of Sciences three years before, it was also the first really incontrovertible proof of dedifferentiation and metaplasia.

These were hardly new ideas, of course. Dedifferentiation had often been proposed during the previous four decades as the simplest explanation of blastema formation, and a great deal of evidence for it had been amassed. Elizabeth Hay had even published an electron microscope photograph of a blastema cell that hadn't despecialized completely and still contained a piece of muscle fiber. Nevertheless, the idea was dismissed by most of the biologists who wielded influence in grant review committees and universities.

Today, however, dedifferentiation is no longer a dirty word. In part, this is because Dave and I devised a way to produce it artificially, which could be repeated by anyone who cared to. Art Pilla, an electrochemist working with Andy Bassett in New York, was the first to confirm our method. I'm happy to have been able to play a major role in this hard-won advance of knowledge.

Even more important, this was the first work of mine that led directly to a technique that helped patients—electrical stimulation of bone healing (see Chapter 8). Meanwhile, our results led to another major question: Couldn't the currents we'd found be used artificially to stimulate other types of regeneration? We decided to see if we could bring limb regrowth a step closer to humans by trying to induce it in rats.

Seven

Good News for Mammals

Stephen D. Smith was the first to induce artificial regrowth with electricity applied to the limb of a nonregenerating animal. In 1967 Smith, setting forth on his own at the University of Kentucky after his apprenticeship to Meryl Rose at Tulane, implanted tiny batteries in adult frogs' leg stumps. I followed his work eagerly and was elated to hear that he'd gotten the same amount of partial regrowth that had resulted from Rose's salt, Polezhaev's needles, and Singer's rerouted nerves. Of all the experiments that have influenced me, this was probably the one that encouraged me the most.

For a battery that was small and weak enough, Smith had returned to the simple technology of Galvani and Volta. He soldered a short piece of silver wire to an equal length of platinum wire, and put some silicone insulation around the solder joint. He chose these two metals as being the least likely to release ions and produce spurious effects by reacting with the surrounding tissue. When immersed in a frog's slightly saline body fluids, this bimetallic device produced a tiny current whose voltage was positive at the silver end and negative at the platinum end.

Since our work on frog erythrocytes hadn't yet been published, it was sheer luck that the current from these batteries fell close to the "window of effectiveness" for blastema formation. As Smith later wrote: "It would be nice to be able to say that I had worked out all the parameters in advance, and knew exactly what I was doing, but such was not the case. As so often has happened in the history of science, I stumbled onto the right procedure."

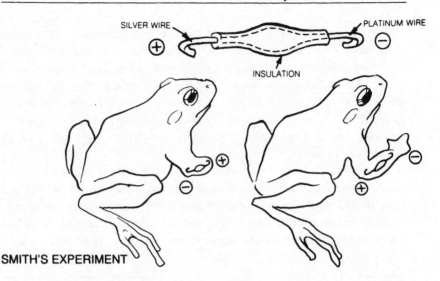

SILVER WIRE   PLATINUM WIRE   INSULATION

SMITH'S EXPERIMENT

Smith implanted his wires along the bone remnant, with one end bent over into the marrow cavity. The limbs with the positive silver electrode at the cut showed no growth, and in some cases tissue actually disintegrated. The negative platinum ends, however, started regeneration; the new limbs all stopped growing at about the same distance from the device, suggesting that regeneration might have been complete if the batteries had been able to follow along. In 1974 Smith made a device that could do just that, and achieved full regrowth.

Despite Smith's success, there was no reason to suppose that his method would work in mammals. One researcher had recently noted some regeneration in the hind legs of newborn opossums, but, since marsupials are born very immature and develop in the mother's pouch to a second birth, we suspected that this was merely a case of embryonic regrowth. Most fetal tissues were known to have some regenerative ability while they weren't yet fully differentiated. Richard Goss had shown that the yearly regrowth of deer and elk antlers was true multitissue regeneration, but this feat seemed too specialized to make us confident about restitution in other mammals or other parts of the body.

Many thought all such attempts were doomed, because the process of encephalization had progressed much further in mammals than amphibians. All vertebrates were known to have roughly the same ratio of nerve tissue to other kinds of tissue, but in mammals most of the limited nerve supply went into the ever more complex brain, until, as Singer had shown in a recent study, the proportion of nerve to other tissue in rat legs was 80 percent less than in salamander legs. This was well below the critical mass needed for normal regeneration, and we thought it might be impossible to make up the difference artificially.

Even if we could supply the proper electrical stimulus, we weren't sure there would be any cells able to respond to it. Mammalian red blood cells had no nuclei, so they couldn't dedifferentiate. Based on our work on bone healing in frogs, we suspected that immature red corpuscles in the bone marrow might take over, but perhaps they were programmed to dedifferentiate only for fracture healing. Even if they would respond to an external current, we wondered whether there were enough of them to do the job.

There was also the problem of complexity. Many regeneration researchers believed that mammalian tissues had become so specialized and complicated that they'd simply outgrown the control system. Maybe it couldn't handle enough data to fully describe the parts needed. If so, any blastema we produced would just sit there, not knowing what to make.

A First Step with a Rat Leg

I tested the kind of silver-platinum couplings Smith used and found they delivered several times too much current for ideal dedifferentiation, according to our frog experiments. Joe Spadaro, another of Charlie's grad students, suggested that we put carbon resistors between the two metals, giving us devices of various current levels.

In 1971, Joe and I amputated the right forelegs of thirty-five rats. We made the cuts through the upper foreleg well away from the elbow so that only the bone shaft, which had long ago ceased growing, would remain at the tip. We used all males, to obviate as many hormonal variables as possible. As controls we treated some of the stumps with no device, or one made of a single metal, or one with the silver positive end facing the stump. We did the actual test on twenty-two of the rats, implanting our batteries with the negative platinum electrode at the wound. We tucked the outer electrode into the marrow cavity and sutured the inner one to the skin of the shoulder.

We had an answer fast. After three days the stumps of the controls had begun to heal over or even, in the case of the highest-current couplings, die back a little behind the amputation line. But the experimental legs with our medium-current devices, supplying 1 nanoamp, were doing well. In a week, nearly every one had a well-formed blastema and seemed ready to replace the whole limb.

Since healing is very fast in rats, and because we wanted a uniform sample for our first test, we sacrificed all of the controls and most of the test animals at this time, although we spared a few for a month. We cut off the entire healing limb, then fixed, stained, and sectioned it for the microscope.

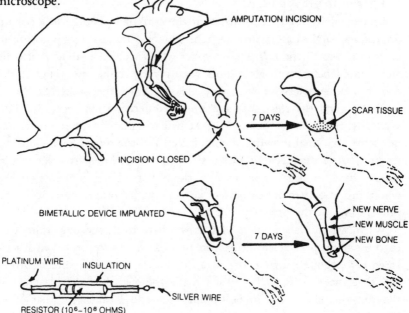

DC STARTS LIMB REGENERATION IN RATS

I shall never forget looking at the first batch of specimens. The rat had regrown a shaft of bone extending from the severed humerus. At the proper length to complete the original bone there was a typical transverse growth plate of cartilage, its complex anatomical structure perfectly regular. Beyond that was a fine-looking epiphysis, the articulated knob at each end of a limb bone. Along the shaft were newly forming muscles, blood vessels, and nerves. At least ten different kinds of cells had differentiated out from the blastema, and we'd succeeded in getting regeneration from a mammal to the same extent as Rose, Polezhaev, Singer, and Smith had done in frogs.

Slides from some of the other animals were even more spectacular. One stump had two cartilaginous deposits that looked like precursors of the two lower arm bones beyond a fully formed elbow joint. All of the regenerates were bent toward the electrode, and in one the lower humerus had formed alongside the old shaft rather than as an extension of it, but otherwise its structure was quite normal.

With one exception, slides from longer than a week were less exciting. They seemed to have gotten *less* organized as time went on. Behind one of these older blastemas, however, at the end of a nearly unformed ghost of a bone, we found cartilage in a five-fingered shape—this limb had begun to grow a hand.

In general, though, it looked as if the current had to be of a certain duration as well as a certain strength. This was no less disappointing to us than it was to the *Life* photographer who visited the lab at that time and wanted before and after shots with a rat playing the piano at the end, but nonetheless we were very pleased. Since the blastemas always formed around the electrodes and since redifferentiation proceeded into organized tissue, we knew the current had stimulated true regeneration, not some abnormal growth. Mammals still had the means for the orderly reading out of their genetic instructions to replace lost parts. We would simply have to learn more exactly the electrical requirements of the whole process, then make devices to supply the proper current at the proper time in the proper place.

When we published our results, it was hard to shroud our excitement in the circumspect scientific jargon needed. We wrote that we'd activated true, though partial, regeneration with a minuscule direct current and that the marrow cells seemed to be the source of the blastema. I thought this claim was sober enough. Joe and I cautioned that other factors had yet to be studied. Most important, we warned that if such a tiny force could so easily switch on growth, it must be very powerful, and we'd best know it thoroughly before using it routinely on humans, lest we give them unwelcome growths—tumors.

I felt that, within the constraints of scientific propriety, we'd uttered a rousing call for a big research push to open up the benefits of regeneration to humans. It must have been a whisper, though, for it caused no more ripples than a feather settling on a frog pond.

Philip Person, a dental surgeon at the Brooklyn VA hospital and a friend whom I'd known for years, asked me to present our results to the New York Academy of Medicine. Before the academy would permit this, however, it insisted that two experts must visit the lab and look at the actual data. One was Marc Singer, who enthusiastically agreed that we'd really started regeneration in the rat. The other man was totally negative, but he wasn't a specialist in regeneration, so the academy permitted me to speak.

Singer was one of the few who showed much enthusiasm when I'd finished reading my paper at the meeting. Most of the audience was unresponsive; there were few comments or criticisms. To these people, electric growth control was still a vitalistic impossibility, and they seemed unwilling to discuss dedifferentiation. The man who'd visited our lab with Singer complained that the amount of new growth was small. Phil pointed out that it wasn't the quantity but the quality of new tissue that was important, especially in such a short time. Singer, convinced of the paramount importance of the nerves, thought the current might be stimulating them rather than directly causing dedifferentiation, but still thought the experiment was a big step forward. Nevertheless, it wasn't even attempted again until seven years later, when Phil Person himself took on the task; he, and later Steve Smith, confirmed our findings with even better results.

Meanwhile, buried in the literature we found reports that others had *already* observed some regeneration in mammals. In 1934 Hans Selye, the famous researcher into the effects of stress, discovered that a rat's limbs could partially regenerate of their own accord when the animal was two to five days old. Five years later Rudolph F. Nunnemacher of Harvard confirmed Selye's observation. Nunnemacher, however, ascribed the growth to a remnant of the epiphyseal plate. The growth-plate cells, he thought, simply might have kept on growing as normal in the adolescent animal. Selye replied that he'd specifically made sure to amputate the limbs high enough to get all of the epiphyseal plate so he could be certain that any growth was regenerative.

Thus Joe and I found that we'd really just extended the age limit for regeneration in the rat. Indeed, two years later Phil Person showed that even the young adult rats we'd used occasionally exhibited some regrowth, a fact that had puzzled us in a couple of our control animals. So, to be exact, our electrodes had temporarily but drastically boosted the efficiency of the process as it normally waned with age in the rodent. Still, it was the first time that had ever been done in a mammal.

Childhood Powers, Adult Prospects

The amputation of a fingertip—by a car door, lawn mower, electric fan, or whatever—is one of the most common childhood injuries. The standard treatment is to smooth the exposed bone and stitch the skin closed, or, if the digit has been retrieved and was cleanly cut, to try to reattach it by microsurgery. The sad fact is that even the most painstaking surgery gives less than optimal results. The nails are usually deformed or missing, and the fingers are too short and often painful, with a diminished or absent sense of touch.

In the early 1970s at the emergency room of Sheffield Children's Hospital in England, one youngster with such an injury benefited from a clerical mixup. The attending physician dressed the wound, but the customary referral to a surgeon for closure was never made. When the error was caught a few days later, surgeon Cynthia Illingworth noticed that the fingertip was regenerating! She merely watched nature take its course.

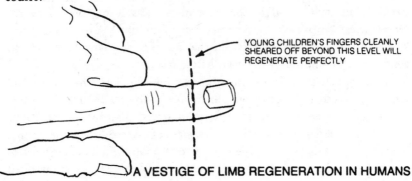

YOUNG CHILDREN'S FINGERS CLEANLY SHEARED OFF BEYOND THIS LEVEL WILL REGENERATE PERFECTLY

A VESTIGE OF LIMB REGENERATION IN HUMANS

Illingworth began treating other children with such "neglect," and by 1974 she'd documented several hundred regrown fingertips, all in children eleven years old or younger. Other clinical studies have since confirmed that young children's fingers cleanly sheared off beyond the outermost crease of the outermost joint will *invariably* regrow perfectly in about three months. This crease seems to be a sharp dividing line, with no intermediate zone between perfect restoration and none at all.

Some pediatric surgeons, like Michael Bleicher of New York's Mount Sinai Hospital, have become so confident in the infallibility of the process that they'll finish amputating a fingertip that's just hanging by a bit of flesh. A lost one will regenerate as good as new, whereas one that has merely been mutilated will heal as a stump or with heavy scarring.

Fingertip regrowth is true multitissue regeneration. A blastema appears and redifferentiates into bone, cartilage, tendon, blood vessels, skin, nail, cuticle, fingerprint, motor nerve, and the half-dozen specialized sensory-nerve endings in the skin. Like limb regeneration in salamanders, this process only occurs if the wound isn't covered by a flap of skin, as in the usual surgical treatment. Illingworth and her coworker Anthony Barker have since measured a negative current of injury leaving the stump.

Sadly, natural replacement has been accepted only at a few hospitals. Bleicher laments his colleagues' resistance to the evidence: "Mention it to young residents just out of the training program, and they look at you as though you're crazy. Describe it on grand rounds or at other institutions, and they tell you it's hogwash." Nearly all surgeons cling instead to flashier and vastly more expensive yet less effective microsurgical techniques, or simple stitches and stunted fingers.

This discovery and our own research indicated that the potential for at least some artificial regeneration was clearly quite good in young mammals. But what about the ones who needed it most—us older folks whose parts were more likely to be injured or broken down? The answer came unexpectedly several years later, in a way that showed the futility of adhering too rigidly to one's original plan. The scientist must be free to follow unexpected paths as they appear.

I always expected each of my associates, whether student or established researcher, to follow some independent project unrelated to our work together. In 1979, a young assistant named James Cullen (now a Ph.D. investigator in anatomy at the Syracuse VA hospital) proposed to study what would happen if nerves were implanted into the bone marrow of rats. Jim thought the nerves should induce new bone to form in the marrow cavity. Since the idea seemed logical and the technique might supplement the electrical bone-healing devices we'd developed by then, I encouraged him to go ahead.

Jim ran into technical problems right away. He could easily dissect the rat's sciatic nerve out of the hind leg, but getting it into the marrow cavity through a hole drilled in the thighbone was like trying to push a strand of limp spaghetti through a keyhole. He resorted to drilling two holes in the femur, passing a wire suture into the outer one, up the femur, and out the hole nearer the hip. Then he looped the wire around the nerve and pulled it into the marrow cavity using the suture. However, after doing a number of these, Jim decided that there had to be a better way. He decided to amputate the rat's hind leg halfway between the hip and the knee. He could then drill a hole into the marrow cavity just below the hip, pass a suture through it, and pull the nerve down the cavity and out the end of the bone remnant. This was much easier and made a better connection of nerve to bone, so Jim prepared a number of animals this way, only to find that the nerve had a disconcerting tendency to pull back, out of the femur. The amputation didn't faze the rats; they used the stump vigorously, and this caused the nerve to retract.

In those few animals whose nerve had stayed in place, an interesting bone formation *had* appeared in the marrow cavity. To secure the nerve and look for the same result in other animals, Jim sutured the nerve to the skin that we closed back over the stump. The stitch held the nerve in place, all right, but one animal so treated gave us a totally unpredicted and fascinating result: The missing portion of the femur partially regenerated. While this was surprising enough, the most startling fact was that Jim had used a group of surplus rats about six months old. These rats were well into adulthood, when mammals were thought to lose all powers of regeneration except fracture healing. What had happened?

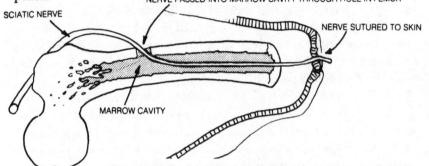

AN ARTIFICIAL NEUROEPIDERMAL JUNCTION IN RATS

Closer examination revealed that we'd made a hole in the skin when we sutured the nerve to it. The nerve appeared to have grown into contact with the epidermis. One of the requirements for normal regeneration of a salamander limb was a neuroepidermal junction, and it looked as though this had formed spontaneously in our one lucky rat when the two tissues were brought together by surgery.

We changed the course of the experiment by operating on the other rats to unite the sciatic nerve and epidermis, after scraping away the dermis. We used animals of various ages. The results exceeded our expectations. Even the old rats regenerated their thighbones and much of the surrounding tissue.

This offered an unparalleled opportunity to find out what it was about the neuroepidermal junction that was so important. We prepared one group of animals with a surgical neuroepidermal junction exactly as before. We prepared a second group the same way, except that we sutured the nerve to the end of the bone, a millimeter away from the hole and with no contact with the epidermis. The first group regenerated, while the second group showed normal rat healing with no growth. The important observation, however, came from electrical measurements we made every day on the stumps. In those animals that formed a neuroepidermal junction, we found electrical potentials following the same curve I'd found in the salamander. The voltage was about ten times as high, but the pattern was exactly the same. In the animals having no neuroepidermal junction, the potentials followed the same curve as in the nonregenerating frog.

We'd discovered that the specific electrical activity that started regeneration was produced by the neuroepidermal junction, not by the simple bulk of nerve in the limb. My original idea that the direct-current control system was located in the nerves now had to be expanded to include the electrical properties of the epidermis as well. The nerve fibers joined the epidermal cells like plugs into sockets to complete the exact circuit needed for a dedifferentiative current. Furthermore, since the neuroepidermal junction was located over the end of the stump, it continually produced blastemal cells exactly where needed, at the growing tip. This discovery was enormously important, then, because it proved beyond doubt that the electrical current was the primary stimulus that began the regenerative process, and that it could operate even in mammals.

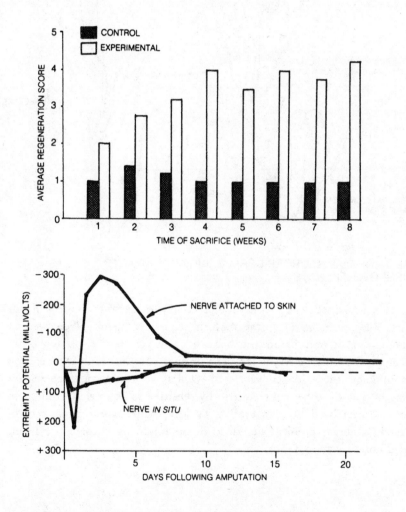

PRODUCING A REGENERATION CURRENT IN RATS

Another experiment showed us a formidable obstacle, however. In an additional series of rats we made a neuroepidermal junction, not at the end of the amputation stump, but on the side of the leg. There we measured the same "regenerative" electrical changes, but nothing happened. There was no growth. This meant that there were no sensitive cells at this site—no cells able to dedifferentiate in response to the current. In mammals, it seemed, such cells were found only in the bone marrow, a sparse cell population to serve as a source of raw material, especially in adult animals.

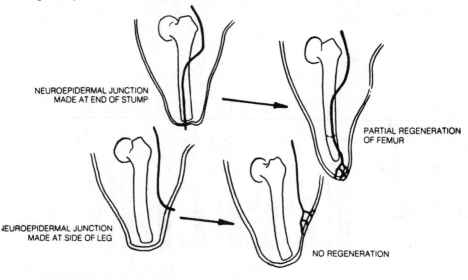

NEUROEPIDERMAL JUNCTION
MADE AT END OF STUMP

PARTIAL REGENERATION
OF FEMUR

NEUROEPIDERMAL JUNCTION
MADE AT SIDE OF LEG

NO REGENERATION

AN ARTIFICIAL NEUROEPIDERMAL JUNCTION STARTS REGROWTH IN RATS

This explained why we never got complete regrowth in any of our rats. The results were typical of an inadequate blastema. There weren't enough sensitive cells in the bone marrow to make a blastema big enough to produce a whole leg. The prospects for full limb regeneration in humans, then, looked very dim—unless we could come up with a way of making other cells electrically sensitive so as to transform them into despecialized blastemal cells. Luckily, while working on a completely different problem described in the following chapter, we stumbled upon a way to do just that.

Part 3

Our Hidden Healing Energy

Disease [is] not an entity, but a fluctuating condition of
the patient's body, a battle between the substance of
disease and the natural self-healing tendency of the body.

—HIPPOCRATES

Eight

The Silver Wand

When Apollo whisked Aeneas off the field of battle before Troy, he healed the hero's shattered thighbone in a matter of minutes. Without a god at the bedside, the process takes three to six months, and sometimes it fails. If the bones didn't knit, the limb formerly had to be amputated after the victim had suffered for a year or more.

It was only in 1972 that I felt ready to try electrical stimulation of human bone growth in such cases. Zachary B. (Burt) Friedenberg, Carl Brighton, and their research group at the University of Pennsylvania had already reported the first successful electrical healing of a nonunion two years before, but to avoid possible side effects we felt we must duplicate the natural signal more closely than they had, and we didn't know enough until after our work on rat leg regeneration. Like Friedenberg we decided to place a negative electrode between the bone pieces, but using a much smaller current and a silver electrode rather than stainless steel. We thought silver would be less likely to react chemically with the tissue and better able to transmit the electrical current. At that time we were treating a patient whose condition seemed to demand that we try the new procedure.

Minus for Growth, Plus for Infection

Jim was in bad shape. Drafted during the Vietnam War, he'd been a reluctant, rebellious soldier. He survived his tour in Nam and was transferred to an Army base in Kansas late in 1970. On New Year's Eve he broke both legs in an auto crash. The local hospital put him in traction, with pins drilled through the skin and bones to hold the pieces together. When he was moved to the base hospital a few days later, all the pins had to be removed due to infection.

Jim's doctors couldn't operate because of the bacteria, so they had to be satisfied with a cast. Because he'd broken one leg below and the other above the knee, he needed a huge cast called a double hip spica. He was totally encased in plaster, from his feet to the middle of his chest, for six months. By August, his left lower leg had healed, but the right femur showed no progress at all. The quarter-inch holes where the pins had been were still draining pus, preventing surgery. That September he was given a medical discharge and flown to the Syracuse VA hospital.

When I first saw him, he was still in a large cast, although now his left leg was free. The halves of the right thighbone were completely loose. There was nothing in standard practice to do but leave the cast on and hope. After six more months Jim's hope was just about gone. For a year he'd lain in bed, unable to leave the hospital for even a brief visit home. He vented his rage against the staff, then grew despondent and unable to face the future, which no longer seemed to include his right leg.

Then Sal Barranco, a young orthopedic surgeon in his last year of residency, was assigned to my service from the medical school. He'd already been a good doctor when he briefly worked with me two years before—smart, hardworking, and really interested in his patients. He took over Jim's care, spent many hours talking with him, and arranged for counseling. Nothing seemed to help. Jim slipped further and further away from us.

Sal had always been interested in what was going on in the lab. In fact, I'd often tried to interest him in a career of teaching and research, but he preferred surgery and its rewards of helping people directly. In February of 1972, as we were nearing the clinical stage with our bone stimulator, Sal said, "You know, Dr. Becker, you really should consider electrically stimulating Jim's fracture. I don't see anything else left. It's his last chance."

The problem was that none of Friedenberg's patients had been infected. Although Jim's septic pin tracts weren't right at the fracture, they were too close for comfort. If I stirred up those bacteria when I operated to insert the electrodes, the game was lost. Moreover, it was obvious by now that electricity was the most important growth stimulus to cells. Even if it produced healing, no one could be sure what these cells would do in the future. They might become hypersensitive to other stimuli and start growing malignantly later. This was the first time in the history of medicine that we could start at least one type of growth at will. I was afraid of beginning a clinical program that might seize the public's fancy and be applied on a large scale before we knew enough about the technique. If disastrous side effects showed up later, we could lose momentum toward a revolutionary advance in medicine. I decided that if I carefully explained what we proposed to do, with all its uncertainties, and let the patient choose, then ethically I'd be doing the right thing.

As to the infection, for several years we had been looking for a way to stop growth. My experiments with Bassett on dogs back in 1964 suggested that just as we could turn growth on with negative electricity, so we could turn it off with positive current. If true, this obviously could be of great importance in cancer treatment. Because ours was always a needy program, trying to do more than we had grants for, we couldn't afford the expensive equipment needed to test the idea on cancer cells. We had to settle for bacteria.

In preliminary tests we found that silver electrodes, when made electrically positive, would kill all types of bacteria in a zone about a half inch in diameter, apparently because of positive silver ions driven into the culture by the applied voltage. This was an exciting discovery, because no single antibiotic worked against all types of bacteria. I thought that if I inserted the silver wire into Jim's nonunion and the area became infected, I could as a last resort make the electrode positive and perhaps save the leg a while longer. Of course, the positive current could well delay healing further or actually destroy more bone.

I explained all this to Jim and said that, if he wished, I would do it. I wanted him to know the procedure was untested and potentially dangerous. With tears in his eyes, he begged, "Please try, Dr. Becker. I want my leg."

Two days later, Sal and I operated through a hole in the cast. The fracture was completely loose, with not one sign of healing. We removed a little scar tissue from the bone and implanted the electrode. The part in between the bone ends was bare wire; the rest, running through the muscles and out of the skin, was insulated so as to deliver the minuscule negative current only to the bone.

The infection didn't spread, and Jim's spirits improved. As I made my daily rounds three weeks later, he said, "I'm sure it's healing. I just know it!" I was still nervous when, six weeks after surgery, it was time to pull out the electrode, remove the cast, and get an X ray. I needn't have worried. Not only did the X ray show a lot of new bone, but when I examined the leg myself, I could no longer move the fracture! We put

Jim in a walking cast, and he left the hospital for the first time in sixteen months. In another six weeks the fracture had healed enough for us to remove the cast, and Jim started rehabilitation for his knee, which had stiffened from disuse.

All the pin tracts, especially the ones nearest the break, were still draining, and Jim asked, "Why not use the silver wire on this hole to kill the infection? Then I'll be all done and won't have to worry anymore about infecting the rest of the bone." I had to agree with his logic. If the hole through the muscle to the outside healed shut, the infection would be more likely to spread within the bone. However, I told him that the positive current might prevent the hole from filling in with bone, making a permanent weak spot there.

We put in the electrode and used the same current as before, except reversing its polarity. I had no idea how long to let it run, so I arbitrarily pulled it out after one week. Nothing much seemed to have happened. The drainage might have been a little less, though not much; but I was afraid to use the positive current anymore for fear of further weakening the bone.

Jim left the hospital and didn't keep his next appointment in the clinic. A year later he returned unannounced saying he was just traveling through Syracuse and thought I would like to see how he was doing. He was walking normally, with no pain, placing full weight on the right leg. He said the drainage had stopped a week after he left the hospital and had never recurred. X rays showed the fracture solidly healed and the one pin tract I'd treated filling in with new bone. The pin site on the other leg was still infected, and I said we could treat that in a few days, since we'd improved our technique in the meantime. "No, I have to be moving on," Jim replied. "I don't have a job. I don't know what I'm going to do, but I know I don't want to spend any more time in hospitals."

Sal had been graduated from the residency program a few months after Jim was discharged in 1973, but before he left he spent all his free time in the lab helping us test the bactericidal (bacteria-killing) electrodes. A few previous reports had mentioned inconsistent antibacterial effects, some with alternating current, some with negative DC using stainless steel, but there had been no systematic study of the subject. We tried silver, platinum, gold, stainless steel, and copper electrodes, using a wide range of currents, on four disparate kinds of bacteria, including *Staphylococcus aureus,* one of the commonest and most troublesome.

Soon we were able to explain the earlier inconsistencies: All five metals stopped growth of all the bacteria at both poles, as long as we used high currents. Unfortunately, high currents also produced toxic effects—chemical changes in the medium, gas formation, and corrosion—with all but the silver electrodes. Apparently such currents through most metals "worked" by poisoning both bacteria and nearby tissues.

Our preliminary observations turned out to be right. Silver at the positive pole killed or deactivated every type of bacteria without side effects, even with very low currents. We also tried the silver wires on bacteria grown in cultures of mouse connective tissue and bone marrow, and the ions wiped out the bacteria without affecting the living mouse cells. We were certain it was the silver ions that did the job, rather than the current, when we found that the silver-impregnated culture medium killed new bacteria placed in it even after the current was switched off. The only other metal that had any effect was gold; it worked against *Staphylococcus,* but not nearly as well as silver.

Of course, the germ-killing action of silver had been known for some time. At the turn of the century, silver foil was considered the best infection-preventive dressing for wounds. Writing in 1913, the eminent surgeon William Stewart Halsted referred to the centuries-old practice of putting silver wire in wounds, then said of the foil: "I know of nothing which could quite take its place, nor have I known any one to abandon it who had thoroughly familiarized himself with the technic of its employment."

With the advent of better infection-fighting drugs, silver fell out of favor, because its ions bind avidly to proteins and thus don't penetrate tissue beyond the very surface. A few silver compounds still have specialized uses in some eye, nose, and throat infections, and the Soviets use silver ions to sterilize recycled water aboard their space stations, but for the most part medicine has abandoned the metal. Electrified silver offers several advantages over previous forms, however. There are no other ions besides silver to burden the tissues. The current "injects" or drives the silver ions further than simple diffusion can. Moreover, it's especially well suited for use against several kinds of bacteria simultaneously. It kills even antibiotic-resistant strains, and also works on fungus infections.

For treating wounds, however, there was one big problem with the technique. Its effect was still too local, extending only about a quarter inch from the wire. For large areas we needed something like a piece of window screen made of silver, but this would have been expensive and also too stiff to mold into the contours of a wound.

We'd been doing our clinical experiments with financial support from a multinational medical-equipment company that made our "black boxes," the battery packs with all their circuitry that powered our electrodes. I discussed the problem with the company's young research director, Jack TerBeek, and a few weeks later he came back with a fascinating material. NASA needed an electrically conductive fabric, and a small manufacturing company had produced nylon parachute cloth coated with silver. It could be cut to any size and was eminently flexible.

It performed beautifully. Although the silver ions still didn't get more than a quarter of an inch from it, we could use it to cover a large area. Hopeful that we might have a cure for two of an orthopedist's worst nightmares—nonunion and osteomyelitis (bone infection)—we studied the positive silver technique in the lab and continued to use the negative electrodes to stimulate bone growth in selected patients. Word spread via newspaper and TV reports. We began getting patients from all over the nation, but we didn't accept many for the experimental program due to my conservative viewpoint. I applied the same criterion as before: Electrical treatment had to be the patient's last chance.

While slowly gaining experience, we surveyed the literature to stay informed about other people's work. As of 1976, fourteen research groups had used bone stimulators on some seven hundred patients, for spinal fusions and fresh fractures as well as nonunions, all with seemingly good results.

We'd used our electrical generator on only thirteen patients by then. We were the only ones using silver electrodes, a lucky choice as it turned out; all the others were using stainless steel, platinum, or titanium. We used 100 to 200 nanoamps per centimeter of electrode, while Brighton and most other investigators were using 10,000 to 20,000 nanoamps. The low level approximated the natural current and also minimized the chances of a dangerous side effect. Brighton and Friedenberg had found a danger of infection and tissue irritation when running their high-current electrodes at more than 1 volt. We figured this couldn't happen at our amperage, but just to be sure we built in an alarm circuit to shut off our box automatically if the electrical force rose close to 1 volt.

By this time we'd also cleared up several more cases of osteomyelitis by reversing the battery and making the silver electrode positive for a day. It looked safe. There was no crossover of effects: When negative, the wire didn't make infectious bacteria grow, and when positive, it didn't destroy bone-forming cells or prevent them from growing when we switched the current to negative. Our confidence in this method grew with one of our most challenging cases, which also forced us to revise our theories.

Positive Surprises

In December of 1976 a young man was sent to our clinic for a possible amputation. John was a man of the north woods. Weathered and hard, he faced the problem philosophically. "What's got to be has got to be," he said through tight lips. Three years before, he'd been in a snow-mobile accident, breaking his right tibia (shinbone) in three places and also fracturing the fibula, the smaller bone of the lower leg. He'd been treated at a small local hospital, where the broken bones had become infected. He'd undergone several operations to remove dead bone and treat the infection, but the bacteria continued to spread. He came to us with the fracture still not healed and with a long cavity on the front of his shin in which one could see right into the dead and infected bone. He was struggling to walk in a cast extending up to his hip. He was married, with five young children, and his leg was obviously not the only place he was having trouble making ends meet.

"What kind of work do you do?" I asked him.

"I trap muskrats, Doc."

"That's all?"

"That's all, Doc."

"How in hell do you manage with that cast on?"

"I put a rubber hip boot over the cast, Doc."

Muskrat trapping is hard work, a tough way to make a living even for a man with two good legs. "John, if you have an amputation and wear an artificial leg, you sure won't be able to do that. What will you do then?"

"I dunno—welfare prob'ly. Prob'ly go nuts."

"You really like to work in the woods, don't you?"

"Wouldn't do anything else, Doc."

"Well, let's get you admitted to the hospital. Something has to be done, and I have an idea that might let you keep your leg." For the first time, John smiled.

In fighting the infection, the first step was to identify the enemy, the microbes. John's wound was a veritable zoo. There were at least five different types of bacteria living in it. Even with only one kind, os-teomyelitis is notoriously hard to treat. Very little blood reaches the bone cells, so both antibiotics and the body's own defense agents are hampered in getting where they're needed. And even if we could get it into the bone, no single antibiotic could fight all of John's germs. Even a mixture would probably create a greater problem than it solved, for any bacteria resistant to the mix would spread like wildfire when the others competing against them were killed.

John's X rays were as chaotic as his bacteria cultures—pieces of dead bone all over the place with absolutely no healing—but we had to deal with the infection first. Since we'd have to use positive current for quite a while, I was afraid we'd destroy some of the bone, but I told John that six months after we got the wound to heal over with skin, I would bring him back into the hospital and use the negative current to stimulate whatever was left. I couldn't promise anything and, since I hadn't yet tried the silver nylon on this type of wound, we might run into unexpected problems. But John agreed with me that he had nothing to lose except his leg, which would certainly have to come off if we didn't try my plan.

A few days later I debrided the wound, removing the dead tissue and all grossly infected or dead bone. There wasn't much left afterward. It was an enormous excavation running almost from his knee to his ankle. In the operating room we soaked a big piece of silver nylon in saline solution and laid it over the wound. It had been cut with a "tail," serving as the electrical contact and also as a sort of pull tab that we could keep dry, outside of the cavity. We packed the fabric in place with saline-soaked gauze, wrapped the leg, and connected the battery unit.

I watched John anxiously during the first two days. If trouble was going to occur, that was when I expected it. By the third day he was eating well, and the current was beginning to drop off, indicating more resistance at the surface of the wound. Now it was time to change the dressing. We were overjoyed to see that the silver hadn't corroded and the wound looked great. Carefully I took a bacterial culture and applied a new silver nylon dressing.

The next morning Sharon Chapin, an exceptional lab technician who took an active part in some of the research, showed me the bacterial cultures. The number of bacteria had dropped dramatically. I went to give John the good news and change his dressing again, when I realized that I could teach him to do his own daily dressing changes. They were time-consuming for me, but John had too much time on his hands and was the one most interested in doing the best thing for his leg. It was a nice feeling to teach a muskrat trapper, who had dropped out of school at sixteen, how to do an experimental medical procedure. He learned fast, and in a day or so he was changing dressings himself and measuring the current, too. By the end of the week, he allowed as how he did a better job than I did. Maybe he did at that, because by then all of our bacterial cultures were sterile—all five kinds had been killed. The soft healing tissue, called granulation tissue, was spreading out and covering the bone. In two weeks, the whole base of the wound, which had been over eight square inches of raw bone, was covered with this friendly pink carpet. The skin was beginning to grow in, too, so we could forget about the grafts we thought we'd need to do.

I decided to take an X ray to see how much bone he'd lost. I could hardly believe the picture. There was clearly some bone growth! We'd been working through a hole in the cast, so I had no idea if the fracture was still loose. Without telling John why—I didn't want to get his hopes up if I was wrong—I removed the cast, felt the leg, and found that the pieces were all stuck together. John watched, and when I was done he lifted his leg into the air triumphantly. It held straight against gravity. His grin opened broader than an eight-lane highway. "I thought you said the bone wouldn't heal yet, Doc!"

I'd never so much enjoyed being wrong, but I warned John not to get too excited until we were sure the good news would hold up. I put him back into a cast and continued treatment another month, until the skin healed over. By then the X rays showed enough repair to warrant a walking cast. John left the hospital on crutches and promised not to run around in the swamps until I told him it was all right. He didn't come back until two months later. The cast was in tatters, and he walked in without crutches, smiling at everyone. The last X rays confirmed it: Healing was nearly complete, and John went back to the wilds.

By mid-1978 we'd successfully treated fourteen osteomyelitis patients with the positive silver mesh wire. The funny thing was, in five of them we'd healed nonunions as a "side effect," without any negative current at all. Obviously it was time to revise our idea that negative electricity alone fostered growth and positive inhibited it.

Andy Marino, Joe Spadaro, and I talked it over. Reducing the DC stimulation technique to its essentials, all you needed was an electrode that wouldn't react with tissue fluid when it wasn't passing current. Since a negative electrode didn't give off ions, any inert metal, such as stainless steel, platinum, or titanium, would work with that polarity. But we knew from our lab work that the situation was very different at the positive pole, where the current drove charged atoms of the metal into the nearby environment. We decided it must be chemical, not electrical, processes that were preventing the bacterial growth at the positive electrode. In that case, *maybe* polarity was unimportant in growth enhancement. We postulated that, because silver ions were nontoxic to human cells and the electrical aspect was right, we inadvertently grew bone with positive current. This idea turned out to be quite wrong, but we'll get to that story in due time.

Joe, who was always fascinated by the history of science, now found that none of the contemporary research groups had been the first to stimulate bone repair electrically. We'd all been beaten by more than 150 years. Back in 1812, Dr. John Birch of St. Thomas' Hospital in London used electric shocks to heal a nonunion of the tibia. A Dr. Hall of York, Pennsylvania, later used direct current through electroacupuncture needles for the same purpose, and by 1860 Dr. Arthur Garratt of Boston stated in his electrotherapy textbook that, in the few times he'd needed to try it, this method had never failed. Because of the primitive state of electrical science then, we didn't know how much current these doctors had used. However, the polarity didn't seem to matter, and they used gold electrodes, which were nearly as nontoxic at the positive pole as silver.

Realizing that we still didn't know as much as we'd thought about the growth control lock, we continued to ply the silver key. At least seventy patients with bone infections have now had the silver nylon treatment, including twenty at Louisiana State University Medical School in Shreveport, where Andy Marino ended up after the closing of our lab in 1980. In some of our first cases we noticed a discharge exuding from the tissues and sticking to the mesh when we changed the dressings. We thought it was "reactive" exudate—from irritation by the current—until one day when, during a slight delay in the operating room, I sent a sample of it to the pathology lab. It was filled with such a variety of cells that we had to rule out a simple response to irritation. Instead there was a variety of primitive-looking cell types, looking just like the active bone marrow of children. However, the patients were long past that age, and, besides, their marrow cavities were closed off with scar tissue from their unmended and infected bone injuries. We had to consider another source.

The exudate appeared at the same time as the granulation tissue, which is composed mainly of fibroblasts, ubiquitous connective cells forming a major part of most soft tissues. Since the exudate also contained some fibroblasts, we decided to see if the unfamiliar types had arisen by metamorphosis from them.

We set up a series of three-compartment culture dishes and placed a standard colony of isolated, pure-bred mouse fibroblasts in each. In one section we put a positive silver electrode, in one a negative electrode, and in the third a piece of silver wire not connected to anything.

In cells right next to all three wires, the cytoplasm changed to an abnormal texture in response to ions of dissolved silver, which migrated only about a hundredth of a millimeter. There were no other effects in the control or negative-current chambers.

Around the positive poles, however, this region was succeeded by an area of great activity for a distance of 5 millimeters on all sides. While doing their job of holding things together, fibroblasts have a characteristic spiky shape, with long sticky branches extending in all directions. In this region where silver ions had been driven by the current, many of the cells had changed to a static, globular form in which mitosis didn't occur. They seemed to be in suspended animation, floating freely instead of adhering to other cells or the sides of the dishes as usual. Mixed among them were many featureless cells with enlarged nuclei, the end products of dedifferentiation. More and more of the rounded fibroblasts turned into fully despecialized cells as the test progressed. Beyond the 5-millimeter line was a border zone with partial changes, followed by a realm of normal, spiky fibroblasts. Dedifferentiated cells normally divide rapidly, but these didn't, perhaps because they were sitting in a plastic dish far removed from the normal stimuli of an animal's body. Within a day after the current was turned off, the cells clumped together into bits of pseudotissue that looked like the young "bone marrow" we'd seen in the exudate. After two weeks they'd all reverted back to mature fibroblasts, presumably because regular replacement of the nutrient medium had by then washed out all the silver ions.

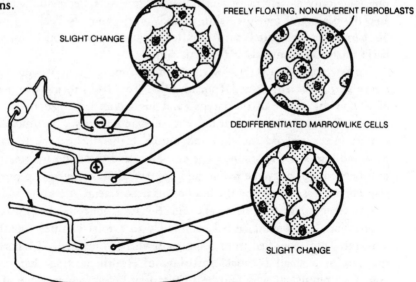

SLIGHT CHANGE

FREELY FLOATING, NONADHERENT FIBROBLASTS

DEDIFFERENTIATED MARROWLIKE CELLS

SLIGHT CHANGE

POSITIVE SILVER DEDIFFERENTIATES CONNECTIVE TISSUE CELLS

To learn more about these astonishing changes, we studied pieces of the granulation tissue itself, taken from patients treated with the silver nylon. We placed the samples in culture dishes and observed them as they grew. Without the silver factor we would have expected a population of slowly proliferating fibroblasts. However, these cells grew fast, producing a diverse and surprising assortment of primitive forms, including fully dedifferentiated cells, rounded fibroblasts, and amoebalike cells. Strangest of all were giant cells that looked almost like fertilized eggs, very active and with several nucleoli. (The nucleolus is a "little nucleus" within the nucleus proper.) When other cells encountered the giant cells, the smaller cells often split open and emptied their nuclei into the giants. After two weeks these diverse cells had coalesced into an amorphous mass of primitive cells closely resembling a blastema, and in another week, as the silver washed out, they'd all become staid, sober fibroblasts acting as though nothing had happened.

The major difference between the two experiments was that the second one started with cells that had already been exposed to positive silver ions *in the human body*. Their rapid growth and unspecialized forms suggested that the fibroblasts in the first experiment had in fact been dedifferentiated. It remains to be seen exactly what the various forms do, but it's obvious that in the aggregate they profoundly stimulate soft-tissue healing in a way that's unlike any known natural process. We ran a controlled study of the healing enhancement on pigs, their skin being physiologically closest to that of humans. Positive silver nylon accelerated the healing of measured skin wounds on the animals' backs by over 50 percent as compared with identical control wounds made on the backs of the same animals at the same time.

We saw positive silver's lifesaving potential most clearly in our experience with a patient named Tom in 1979. Tom had had massive doses of X rays for cancer of the larynx, and his larynx later had to be removed. Because of the radiation, the surrounding tissue was helpless against infection, and the skin and muscle of his entire neck literally dissolved into a horrid wound. The ear, nose, and throat doctor treating him begged me to try the nylon, and I agreed after the attending physician got a release signed by the head of his department. After one month of electrified silver treatment, the infection was gone and healing was progressing, the wound healed completely in a total of three months, although Tom soon died from tumors elsewhere in his body. I reported this case at a small National Institutes of Health meeting that same year. One physician, who said he'd never heard of any comparable healing of such a grave wound, was moved to exclaim after seeing my slides, "I have witnessed a miracle!"

We may only have scratched the surface of positive silver's medical brilliance. Already it's an amazing tool. It stimulates bone-forming cells, cures the most stubborn infections of all kinds of bacteria, and stimulates healing in skin and other soft tissues. We don't know whether the treatment can induce healing in other parts of the body, but the possibility is there, and there may be other marvels latent in this magic caduceus. Just before our research group was disbanded, we studied malignant fibrosarcoma cells (cancerous fibroblasts) and found that electrically injected silver suspended their runaway mitosis. Most important of all, the technique makes it possible to produce large numbers of dedifferentiated cells, overcoming the main problem of mammalian regeneration—the limited number of bone marrow cells that dedifferentiate in response to electrical current alone. Whatever its precise mode of action may be, the electrically generated silver ion can produce enough cells for human blastemas; it has restored my belief that full regeneration of limbs, and perhaps other body parts, can be accomplished in humans.

Many questions remain, however. We don't know how the changed cells speed up healing or how the silver changes them. We don't know how electrically produced silver ions differ from ordinary dissolved ions, only that they do. They evoke widely different reactions from the fibroblasts, and the cells affected by dissolved ions close to the electrode are prevented from dedifferentiating in response to the electrified silver. Most important is a search for possible delayed side effects.

These questions urgently need research by some good electrochemists, but the work isn't being done. We were probably lucky we hadn't found this effect in our first round of lab tests on silver electrodes. The Food and Drug Administration let us test the antibacterial technique on selected patients because we found no toxicity and because a few hours a day was enough. To say that the same electrodes run for a longer time could stimulate healing was such a bold claim that permission probably would have been denied. At this point, however, we sorely need enough imagination on the part of research sponsors to follow up these leads in the laboratories, while making the treatment available now to the desperate few who have no other hope.

The Fracture Market

Where does all this leave us in our understanding of electrically triggered bone healing? I'm afraid we're not too much better off than the nineteenth-century doctors who lost this effective treatment because no one understood it well enough to defend it from electrotherapy's opponents. Of course, we have more pieces of the puzzle today, but we still don't have a complete picture of how any of the competing techniques work.

Besides our low-current silver electrodes, there are two other basic types of bone growth stimulators. Friedenberg and Brighton at first placed their stiff stainless-steel wires through holes drilled into the bone near the break. Now a "semi-invasive" refinement is used in many patients—sticking the electrodes through the flesh into the fracture gap under local anesthetic; several may be needed for a large bone. They're connected to a self-contained battery pack set right into the cast. Friedenberg and Brighton patented their invention, and it's now approved by the FDA and marketed. About three fourths of the groups treating patients use some variation on this theme.

Australian researchers under D. C. Patterson devised a spiral titanium electrode that's placed into a notch cut in the bone on both sides of the fracture. It's now also FDA approved and marketed. Since this device, battery pack and all, must be implanted and removed in two separate operations, and since the electrode usually must be left behind in the bone, late complications may occur.

Others have taken a completely different approach, using pulsed electromagnetic fields (PEMF) to induce currents in the fracture area from outside the body. The best-known proponents of this method, Andrew Bassett and electrochemist Arthur Pilla, worked together at the Orthopedic Research Laboratories of Columbia-Presbyterian Medical Center in New York until 1982; Pilla is now at the Mount Sinai Medical Center. They developed a pair of electromagnetic coils sheathed in plastic pads, connected to a book-sized generator that plugs into a wall socket.

Having experimented with a wide variety of pulsed fields, Bassett and Pilla found four that stimulate fracture healing. The one that works best, which is now also approved and available commercially, is produced by electromagnets driven by alternating current supplied in bursts of pulses. Although it ties the patient to an electrical outlet for twelve hours a day (mostly during sleep, of course), this apparatus completely avoids surgery and its attendant risks.

The funny thing is that all three methods—low current, high current, and PEMF—seem to work equally well. Since the FDA approved them in late 1979, success rates have stabilized at about 80 percent.

As far as the two electrode methods are concerned, I believe some experiments we did in 1977 and 1978 revealed why they *both* work. When we arranged all the reports in order from lowest current to highest, we found a narrow band of low amperages and a wide band of higher ones that worked, separated by a range in between that failed. We tested various nonsilver electrodes on animals and found that, at the effective currents, they all produced some electrochemical changes *even at the negative pole.* Among other products, they created various highly reactive ions called free radicals, essential in cellular chemistry but also destructive in excess. These radicals irritate cells, and, since any continuous irritation stimulates bone-forming cells to lay down new matrix in self-defense, we concluded that the higher-current methods worked primarily by such irritation. Conversely, I believe low-current silver electrodes stimulate bone formation directly—by dedifferentiating the marrow cells and perhaps also by stimulating the periosteal cells.

At first, PEMF research suggested that the coils worked by inducing in the tissues electrical currents that changed the permeability of cell membranes to calcium. In most nonunions at least a small amount of fracture callus has appeared, consisting of collagen fibers, but for some reason it hasn't entered the next stage, in which apatite crystals form on the fibers. Work by Pilla and Bassett suggested that the currents induced by the pulsed fields caused the cells of the callus to absorb large amounts of calcium. Later, when the coils were turned off, they reasoned, the cells dumped this calcium outside among the collagen fibers, and apatite crystals finally formed where they belonged.

Their experiments raised the hope that other wave forms might regulate membrane passage of other ions or even control DNA transcription and protein synthesis. It seemed these field-induced currents might act as "vocal cords," allowing us to "speak" to the cell nucleus via the membrane, much as sound waves communicate with the brain via the ear.

PEMF does in fact induce currents—of a type never found normally in the body. Each pulse produces millions of tiny eddy currents briefly flowing in circles. As the magnetic field expands at the beginning of a pulse, the currents circle in one direction; as it collapses, they reverse. However, the latest research has cast doubt on the idea that these currents affect specific cellular processes. Rather, it seems that artificial time-varying magnetic fields *directly* activate the cells by speeding up their mitotic rate, as discussed in more detail in Chapter 15.

You may ask, "As long as something works, why quibble about how?" My answer is that understanding how is our best hope for using the tool right, without causing our patients other problems later. By sticking our own neologisms into the cell's delicate grammar, we automatically risk garbling it in unforeseen ways.

As of now, we're like blind people crossing a minefield. Accelerated mitosis is a hallmark of malignancy as well as healing, and long-term exposure to time-varying electromagnetic fields has been linked to increased rates of cancer in humans. Bassett has discounted potential dangers, saying, "You would experience almost the same field strength by standing under a fluorescent light." However, a fluorescent lamp may well feel like a floodlight to cells that can see nanoamperes of current. One of the main lessons of bioelectromagnetism so far is that less is often more.

On the other hand, it's too easy to assume that "natural is better." Since it would vindicate our low-current method, I obviously hope it's true, but the fact remains that putting electrodes in a bone is itself a very unnatural act. Stimulation of healing outside the normal limits of the process may incur *fewer* risks in the end. So far the evidence suggests otherwise, but we don't yet know for sure. That's why I keep emphasizing the need to go slowly, using these contraptions only when all else has failed, until we understand them better.

The most urgent need is a search for possible malignant effects. As far as I know, I've done the only such research on electrodes to date—one simple tissue culture study without grant support, using some money I saved from our last research project funded by the manufacturer of the battery pack. I proposed more extensive tests to various granting agencies before our lab was closed, but was turned down every time.

I exposed standard cultures of human fibrosarcoma cells to 360 nanoamps from stainless steel electrodes. I tried it five times, and each time there was a threefold increase in cell population at both electrodes. Even for cancer cells, this is remarkable proliferation for such a short time. To my knowledge, none of the developers or marketers of electrode devices have chosen to duplicate this test or try their own, despite the ease of doing the work and the fact that they have plenty of money. Whatever evidence on this point that may have been presented to the FDA hasn't been made public. After the evaluation panel granted commercial approval, however, several of its members expressed fears that this possibility hadn't been adequately tested. At this time, therefore, I must conclude that high-current electrodes *might* enhance the growth of any preexisting tumor cells in the electrical path—unlike low-current silver, which when negative had no effect on, and when positive suspended, mitosis of cancer cells in our lab.

As for PEMF, Bassett and Pilla believe that only cells in a healing process gone awry can "hear" their wave form, so they discount the idea of cancer enhancement from it. They claim to have found no PEMFs that produce or accelerate malignant growth; on the contrary, Pilla and oncologist Larry Norton of Mount Sinai say they've found at least one that seems to inhibit it in lab animals. This claim is seriously flawed, however, because of the difference between subjecting an entire animal to a magnetic field, and directing the same field to a small area around a fracture (see Chapter 12). Moreover, in 1983, Akamine, a Japanese orthopedic researcher, reported that the pulsed magnetic fields used for bone healing dramatically increased the mitotic rate of cancer cells. The same fields inhibited the return of such "stimulated" cancer cells to a more normal state. Thus PEMF, like high-current treatments, apparently does enhance cancer growth.

In the last decade or so, electrobiologists have learned a great deal about the effects of time-varying electromagnetic fields (as opposed to steady-state fields) on living tissue. We'll review these discoveries in Chapters 14 and 15. The evidence to date indicates that PEMF works by increasing the mitotic rate of the healing cells, not by altering calcium metabolism. If so, it can't possibly discriminate between bone-healing cells and any other type. It will accelerate the growth of *any* cellular system that is actively growing; this includes not only healing tissues, but fetal and malignant tissues as well.

At the present rate of basic research, we won't have direct proof on whether electrical healing stimulators are nurturing seeds of cancer in humans until two or three decades from now. We could find out much sooner by simple experiments on animals having shorter life-spans. Until we have that definitive answer, I contend that all three techniques should remain available as a last resort to prevent loss of limb, but I'm appalled at their increasing use to speed up orthodontics or accelerate healing of fresh fractures.

Unfortunately, the trend is away from caution. By the time this book is published, tens of thousands of patients will have been treated with the devices, many as a first, rather than last, resort. At a recent orthopedic meeting, I learned that four more companies are hoping to market new models. Several have asked me to advise them, but I haven't found one yet that wants to embark on any serious research. Without such a commitment, I refuse to take part in any battle of salesmen. I never even tried to patent the low-current silver method, because a medical device generally isn't considered patentable if the research that went into it was conducted throughout the scientific community and published for all to read. As I see it, the rush to the marketplace can only spawn jurisdictional disputes and ensure that important findings are kept as proprietary secrets.

Electrical osteogenesis could be the opening wedge into a new era of medicine. Within a few years, we may know how to use these tech-

Nine

The Organ Tree

"I have yet to see any problem, however complicated, which when you looked at it the right way did not become still more complicated," science fiction writer Poul Anderson once observed. To a certain extent this is true of regeneration. Intricate nature is still more than a match for our finest-spun hypotheses. Yet we've now reached the oasis of science that we call an interim understanding, where the data begin to shake into place and we can sense the pattern of the rebus from the blanks we've filled in.

Ultimately we must relate all we learn about regeneration to a general system of communication among cells, for regrowth is only a special case of the cooperative cohesion that's the essence of multicellular life. This communication system includes but extends beyond the gene-protein-enzyme subsystems that govern the specialization of cells and unite their chemical trade routes into smoothly working tissues and organs. During embryonic development, cells where muscle will appear must receive instructions from their environment telling them to repress all genes except the muscle genome, or subcode. In many tissues, perhaps in all, chemical inductors from previously formed tissue perform this task, leading embryonic cells step by step through the stages of differentiation. However, chemical reactions and the passage of compounds from cell to cell can't account for structure, such as the alignment of muscle fiber bundles, the proper shape of the whole muscle, and its precise attachment to bones. Molecular dynamics, the simple gradients of diffusion, can't explain anatomy. The control system we're seeking unites all levels of organization, from the idiosyncratic yet regular outline of the whole organism to the precisely engineered traceries of its microstructure. The DNA-RNA apparatus isn't the whole secret of life, but a sort of computer program by which the real secret, the control system, expresses its pattern in terms of living cells.

This pattern is part of what many people mean by the soul, which so many philosophies have tried to explicate. However, most of the proposed answers haven't been connected with the physical world of biology in a way that offered a toehold for experiment. Like many attempts, the latest major scientific guess, the morphogenetic field proposed by Paul Weiss in 1939, was just a restatement of the problem, though a useful one. Weiss conjectured that development was guided by some sort of field projected from the fertilized egg. As the dividing cell mass became an embryo and then an adult, the field changed its shape and somehow led the cells onward.

The problem was that there were too many "somehows." Even if one accepted Burr's largely ignored measurements of an electric L-field and admitted that it might be the morphogenetic field (a possibility Weiss dogmatically rejected), there was still no way of telling where the L-field came from or how it acted upon cells. Nor was there an explanation of how, if the field was an emanation *from* the cells, it could also *guide* them in building an animal or plant. In applying the idea to regeneration, biologists faced the related and seemingly insurmountable problem of how a more or less uniform outflow of energy could carry enough information to characterize a limb or organ. Given the complexity of biological structures, this was even harder than imagining how a field could "somehow" survive when the part it referred to was missing.

However, the morphogenetic field no longer has to account for everything. Acceptance of dedifferentiation lets us divide regrowth into two phases and better understand each. The first phase begins with the cleanup of wound debris by phagocytes (the scavenger race of white blood cells) and culminates in dedifferentiation of tissue to form a blastema. Redifferentiation and orderly growth of the needed part constitute the second phase.

Simplifying the problem in this way should give biologists an immediate sense of accomplishment, for the first stage is now well understood. After phagocytosis, while the other tissues are dying back a short distance behind the amputation line, the epidermal cells divide and migrate over the end of the stump. Then, as this epidermis thickens into an apical cap, nerve fibers grow outward and subdivide to form individual synapselike connections—the neuroepidermal junction (NEJ)—with the cap cells. This connection transmits or generates a simple but highly specific electrical signal in regenerating animals: a few hundred nanoamperes of direct current, initially positive, then changing in the course of a few days to negative.

The pituitary hormone prolactin, the same substance that stimulates milk flow in nursing mothers, seems to sensitize cells to electric current. Then the signal causes nearby cells to dedifferentiate and form a blastema, apparently by changing the way cell membranes pass calcium ions. After confirming our frog blood-cell work, Art Pilla went on to produce the same changes by using pulsed DC to make a wave of calcium ions flowing across the culture dish. Steve Smith then confirmed the importance of calcium by preventing dedifferentiation with a calcium-blocking compound, and restarting it with another substance that enhanced passage of calcium ions. Working together, Smith and Pilla next used the same PEMF wave form now in clinical use to nearly double the rate of salamander limb regeneration, while completely preventing it with a different pulse pattern. Widespread recent work on calcium-binding proteins, such as calmodulin, has made it fairly certain that electrical control of calcium movement through cell membranes directs the give-and-take among these proteins, which in turn supervises the cell's entire genetic and metabolic industry.

Although not conclusive, the available evidence suggests that the current flows through the perineural cells rather than the neurons themselves (see Chapter 13). These are several types of cells that completely surround every nerve cell, enclosing all the peripheral fibers in a sheath and composing 90 percent of the brain. Lizards can replace their tails without the spinal cord, as long as the ependyma, or perineural cells surrounding the cord, remains intact, and ependymal tissue transplanted to leg stumps gives lizards some artificial regeneration there. However, the circuit may shift tissues near the wound, for Elizabeth Hay's electron microscope studies clearly show that the peripheral nerve's Schwann cell sheaths stop just short of the epidermis, and only the naked neuron tips participate in the NEJ. The exact current pathway in this microscopic area remains to be charted.

Not all cells can respond, however, as Jim Cullen and I found in one part of our fortuitous rat-regeneration experiments of 1979. Dedifferentiation occurred only when we passed the deviated sciatic nerve to the epidermis *through the bone marrow.* When we led it through the muscle yet sutured it to the skin in exactly the same way, an NEJ producing the right current appeared, but no blastema and no regrowth. Muscle cells apparently weren't competent to dedifferentiate in the adult rat. The cellular target proved to be just as important as the electrical arrow.

There's still some opposition to parts of this scenario. Among some scientists, the prejudice against electrobiology remains so strong that one otherwise fine recent review doesn't even mention the NEJ or the difference between currents of injury in frogs versus salamanders!

Other objections are a little more substantial. A Purdue University group has measured electrical potentials near the surface of regenerating limbs underwater, using a vibrating probe. This is an electrode whose tip, ending in a tiny platinum ball, oscillates rapidly to and fro, giving the average voltage between the two ends of its motion. These researchers describe an arc of ion flow—they categorically deny the possibility of electron currents in living tissue—out from the stump and through the water or, in semiaquatic animals, a film of moisture on the skin. From the water, they suggest, these ions travel to the limb skin behind the amputation, then in through all the inner tissues, and finally out of the stump again to complete the circuit. They believe the epidermis drives these currents by its normal amphibian function of pumping sodium (positive) ions from the outside water into the body. They conceive of this ion flow as the regeneration current itself, because changing the concentration of sodium in the water directly affects their current measurements, and because certain sodium-blocking techniques have interfered with limb regrowth in about half of their experimental animals.

Of course, the Purdue researchers don't dispute the amply proven need for nerve and injury currents during regeneration. They've even confirmed Smith's induction of leg regrowth in frogs with batteries generating *electron* currents. Nevertheless, they consider nerves the target rather than the source of current, even though they propose no reason why their ion flow should be restricted to nerve tissue. In fact, they base their hypothesis partly on evidence that sodium flows even from denervated limbs.

There are several other problems with this theory: Its proponents' own measurements show that the sodium ion current almost disappears just when its supposed effect, blastema formation, is occurring. Moreover, it fails to explain the easily observed reversal of polarity in injury currents measured directly on the limb, as well as the crucial role of the NEJ. The proposed circuit goes right past the NEJ! Finally, it can't account for the several tests of semiconducting current throughout the nervous system, or regeneration in dry-skinned animals such as lizards.

To the lay person all this may seem like academic hairsplitting, until we reflect on the stakes: understanding regeneration well enough to restore it to ourselves. Certainly skin is electrically active. It's piezoelectric and pyroelectric (turning heat into electricity) as well as a transporter of ions in wet animals. In the last two decades nearly all tissues have been proven to produce or carry various kinds of electrical charge. Skin may play an as-yet-unknown role in regeneration besides its part in the NEJ, or it may merely be producing unrelated electrical effects. In any case, there are far too many data about the role of nerves to call skin the major source of the regeneration current. In fact, even the Purdue group has measured a stump current that's independent of sodium concentration.

A recent experiment by Meryl Rose gave further evidence of neural DC, without clearing up all aspects of the question. Rose removed the nerves from larval salamander legs before amputation. Normally such denervated larval stumps die right back to the body wall, but when Rose artificially supplied direct currents like those I measured in my first experiment, they regrew normally. This is pretty conclusive proof that the nerves are the electrical source in phase one. However, since it looks as though nerves also organize regeneration's second phase (see below), it's hard to understand how the new legs could have been completely normal when they were disconnected from the rest of the nervous system. Perhaps salamanders can pinch-hit for nerves at this stage through a tissue other than nerve. On the other hand, new nerves may simply have regrown into the limbs unbeknownst to Rose by the later stages of the experiment.

Phase two begins as the embryonal cells pile up and the blastema elongates. Early in this stage a sort of spatial memory becomes fixed in the blastema cells so the limb-to-be will have its proper orientation to the rest of the body. At the same time or shortly afterward, the cells at the inner edge of the blastema receive their new marching orders and platoon assignments. Then they redifferentiate and take their places in the new structure.

We can infer two things about the control for this part of the process. Since the blastema forms the right structure *in relation to the whole organism,* the guidance can't be purely local, but must come from a system that likewise pervades the whole body. Furthermore, there are no dedifferentiated cells left over when the work is done; there are just enough and no more. Thus there must be a feedback mechanism between the redifferentiation controls at the body side of the blastema and the NEJ's dedifferentiation stimulus at its outer edge.

A large body of earlier work has shown that the redifferentiation instructions are passed along a tissue arc whose main element is the circuit already established between nerves and epidermis in the first phase. The electrical component persuasively explains how this arc, an update of the morphogenetic field, may work. The direction (polarity) plus the magnitude and force (amperage and voltage) of current could serve as a vector system giving distinct values for every area of the body. The electric field surrounding continuously charged cells and diminishing with the distance from the nerve would provide a third coordinate, giving each cell a slightly different electrical potential. In addition, a magnetic field must exist around the current flow, possibly adding a fourth dimension to the system. Together these values might suffice to pinpoint any cell in the body. The electric and magnetic fields, varying as the current varies with the animal's state of consciousness and health, could move charged molecules wherever they were needed for control of growth or other processes. Since currents and electromagnetic fields affect the cell membrane's "choice" of what ions to absorb, reject, or expel, this system—in concert with the chemical code by which neighboring cells recognize each other—could precisely regulate the activities of every cell. It could express the exact point along the limb at which new growth must start; distinguish between right and left, top and bottom; even explain how totally missing parts, like extirpated bones or all the little bones of wrist and hand, can reappear.

Furthermore, the difference between electrical values at the inner and outer edges of the blastema would lessen as a new limb grew behind it. (Remember that the electrical potential grows increasingly negative toward the end of an intact limb.) The gradual convergence of these two values could constitute a feedback signal perfectly reflecting the number of dedifferentiated cells still needed. Although the results weren't entirely conclusive, perhaps because measurements had to be made under anesthesia, several experiments in the 1950s suggested that such a voltage differential governed restitution of the proper number of segments in earthworms. There was even a surge of positive potential that seemed to indicate when the job was finished.

This is a rich concept, and the details are without doubt more complex than this sketch, but they're all open to experimentation in a way that Weiss's morphogenetic field and Burr's L-field were not. The best part of this two-stage analysis is that it gives us a rationale for trying to foster regeneration after human injuries before we know all the details of the second phase.

The rat limb experiments strongly suggest that mammals lack two crucial requirements for the first phase of regeneration: They don't have the necessary ratio of nerve tissue to total limb tissue, the amount needed to make the dedifferentiation stimulus strong enough; and they lack sufficient sensitive cells to respond to the electrical stimulus and form a big enough blastema. The work on rats pointed the way to defining the proper current, and the ability of electrically injected silver ions to dedifferentiate fibroblasts now gives us a possible method for producing an adequate blastema. We should now be able to supply the requirements for phase one in humans. Once this is done, the body itself can probably take care of phase two, even though we don't understand the process. Fingertip regrowth in children suggests that our bodies still have the ability to redifferentiate the cells and organize the missing part, as long as the electrical stimulus and the supply of sensitive cells are sufficient.

Microsurgeons have performed wonders in reimplanting cleanly severed portions of arms, legs, and fingers, but these limbs are subject to atrophy and obviously can't be grafted if they're too badly mangled or riddled with disease. As one who has performed too many amputations in his time, I find the prospect of being able to give a patient the real thing instead of a prosthesis tremendously exciting. There's a good chance that we'll eventually treat some nongenetic birth defects or old injuries by cutting off the defective part and inducing a normal one to grow. Perhaps, combined with gene splicing, such techniques could even rectify genetic birth defects.

Since no one has yet achieved full regeneration in rats or any other mammal, these dreams won't come true overnight. They aren't chimerical, however. The remaining problems could probably be solved in a decade or two of concerted basic research. Meanwhile, human capacities for repair of certain tissues are greater than most people realize, and there are already promising ways of enhancing some of them.

Cartilage

Fossils show that even the dinosaurs had arthritis, but unfortunately it outlived them. Many varieties have been described, all of which result in destruction of the hyaline (glassy) cartilage that lines the ends of the bones. The remaining cartilage cells try to heal the defect by proliferating and making more cartilage. They're almost never equal to the task, and scar tissue fills the rest of the hole. The result is pain, for scar tissue is too spongy to bear much weight or keep the bones from grinding against each other.

After our success at getting rat legs to partially regenerate, we studied this problem in 1973. We reasoned that, since cartilage was made by only one kind of cell, getting it to regrow would be easier than working with a whole limb.

With orthopedic resident surgeon Bruce Baker of the Upstate Medical Center, we removed the cartilage layer from one side of the femur at the knee in a series of white rabbits. The operation left a circular hole of bare bone about 4 millimeters across. In the experimental animals we implanted silver-platinum couplings like those used on the rats, drilling the platinum end into the defect and tucking the rest along the bone. Most of the control animals filled in the defect with scar tissue along with some inferior fibrous cartilage. About a tenth of them grew a millimeter or two of good hyaline cartilage at the edge of the hole. But sure enough, the rabbits with the implants showed greatly enhanced repair. When we used an improved battery implant with silver wires at each end, we got even better results. Two of the rabbits healed the damage completely with beautiful hyaline cartilage just like the original material.

A few years later, when we were testing various electrode metals, we tried a different approach specifically for rheumatoid arthritis, in which runaway inflammation causes phagocytes to attack healthy cartilage cells. Gold salts taken orally sometimes control this disease but often produce toxic side effects. We figured electrical injection of pure gold directly into the joint with no other ions might work better. To find out, Joe Spadaro and I produced rheumatoid arthritis in the knees of both hind legs in forty rabbits, using a standard experimental procedure. Then we treated one knee in each animal with a positive gold electrode stuck right into the space between the two bones for two hours. Joe did the actual treatments. Then we sacrificed the animals gradually over a period of two months, and I examined both arthritic knees, not knowing until later which had been given gold. During the first two weeks about 70 percent of the treated knees were markedly better than the untreated ones. The improvement fell off to about 40 percent thereafter, suggesting that the treatment must be repeated for continued results.

Obviously, these were only preliminary experiments. However, since an estimated 31 million Americans suffer from arthritis, for which there is no cure yet, I think both avenues should be thoroughly explored as soon as possible.

Skull Bones

Lev Polezhaev has spent his career investigating what might be called the Polezhaev principle—the greater the damage, the better the regrowth. He has found he can often enhance repair by adding homogenates, minces, and extracts of the damaged organs, even though this doesn't augment the current of injury as his needling procedure did.

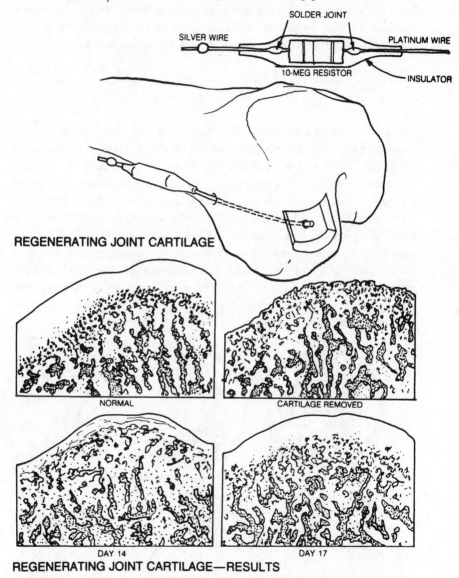

SOLDER JOINT

SILVER WIRE

PLATINUM WIRE

10-MEG RESISTOR

INSULATOR

REGENERATING JOINT CARTILAGE

NORMAL

CARTILAGE REMOVED

DAY 14

DAY 17

REGENERATING JOINT CARTILAGE—RESULTS

Eventually Polezhaev developed a way to induce regeneration of holes in the skull, which normally heal over with scar tissue. As long as the dura mater, the tough membrane between skull and brain, is intact, a paste of blood and fresh (living) powdered bone will induce the bone cells at the edges to grow and bridge the gap. Microscopic studies have shown that the few live cells remaining in the paste don't survive and the bone particles themselves soon dissolve. Instead, some substance from the disintegrated bone stimulates repair. Since its first successful trial on humans by several Russian surgeons in the mid-1960s, this method has gradually come into increasing use in the Soviet Union.

Eyes

There is at present no indication whatsoever that humans could ever regenerate any part of their eyes, but the ability of newts (salamanders of the genus *Triturus*) to do so makes this a tantalizing research ideal for the far future. If the lens in a newt's eye is destroyed, the colored cells of the top half of the iris extrude their pigment granules, then transform by direct metaplasia into lens cells. They soon start synthesizing the clear fibers of which the lens is made, and the whole job is finished in about forty days. In case the iris is gone, too, a newt can create a new one from cells of the pigmented retina, and those cells can also transform into the neural retina layer in front of them. If the optic nerve gets damaged, the neural retina in turn can regenerate the nerve tract backward and reconnect it properly with the brain.

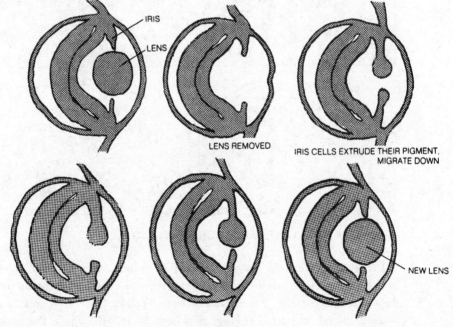

SALAMANDER EYE-LENS REGENERATION

No one knows why newts are so much more adept at this than all other creatures; their eyes have no obvious structural or biochemical peculiarities. Steve Smith gave us an important fact to work with when he found two proteins in the lens that seem to prevent the iris cells from changing into new lens cells as long as the old lens is in place. Since the neural retina must be intact for most of the transformations to happen, it may provide a constant electrical stimulus that goes into effect only when the inhibitory proteins are removed by injury.

No blastema is formed; instead the cells change costume right onstage. Furthermore, certain ingenious experiments have shown that a wound isn't really necessary, only the interruption of the inhibitory mechanism. Therefore, the stimulus from the neural retina probably isn't the familiar injury current of limb regrowth. However, despite a voluminous research literature on newt eye regeneration, no one has yet studied its electrical aspects. This may be why we're still so far from understanding the natural process, let alone trying to adapt it to human eyes.

Muscle

Every muscle fiber is a long tube filled with rows of cells (myocytes) laid end to end with no membranes between them—in effect, one multinucleated cell, called a syncytium. These nuclei direct the manufacture of contractile proteins, which are lined up side by side and visible, when stained, as dark bands across the array of myocytes. Each muscle fiber is surrounded by a sheath, and groups of them are bound together in bundles by thicker sheaths. At the edge of each bundle are long, cylindrical cells with huge nuclei and very little cytoplasm, called myoblasts or spindle cells. Also along the edges, between the spindle cells, clusters of tiny satellite cells can be seen at high magnifications.

After a crushing injury or loss of blood from a deep cut, muscle in the damaged area degenerates. The myocyte nuclei shrivel up and the cells die. Soon phagocytes enter to eat the old fibers and cell remnants. Only the empty sheaths and a few spindle and satellite cells are left.

Now these remaining cells turn into new myocytes, fill up the empty tubes, and begin secreting new contractile proteins. Although the early part of this process proceeds without nerves, it can run to completion only if motor nerve fibers reestablish contact with the terminals, called end plates, that remain at specific distances along each fiber sheath. If these end-plate areas are cut out, the nerve endings will enter, sniff around, and then retract. The muscle will atrophy. If the nerves re-establish the connections, new muscle cells will fill most of the original volume, gradually build up strength, and then completely differentiate into slow-twitch or fast-twitch fibers.

Attempts to enhance muscle regeneration in humans take two approaches. If available, a graft of a whole muscle from another source is the more effective. This is actual single-tissue regeneration, because the original muscle cells die and are replaced after new blood and nerve connections are made. Since its first clinical use in 1971, this approach has proven successful in replacing defective small muscles of the face and also in restoring anal sphincter control. Large limb muscles haven't been successfully transplanted yet.

Another method may soon be used in humans when grafts aren't possible. Muscle regeneration in birds and laboratory mammals has been considerably enhanced by inserting muscle tissue minced with fine scissors into pieces of no more than 1 cubic millimeter. Soviet biologist A. N. Studitsky first devised this method in the 1950s, extending Polezhaev's work, but its development has been slow.

Abdominal Organs

Despite over two hundred years of descriptive work, new regenerative capacities are still discovered in the animal kingdom from time to time. We've recently learned, for example, that adult frogs can restore their bile ducts, although for some reason females are better at it than males. Doctors have long known that the liver can replace most of the mass lost through injury by compensatory hypertrophy, in which its cells both enlarge and increase their rate of division so that the organ's chemical-processing duties can be maintained even though the ruined architecture isn't restored. Similarly, damage in one kidney is made up by enlargement of the other without rebuilding the intricate mazes of microtubules in the glomeruli. Recent studies of rat livers suggest that a combination of insulin and an epidermal growth factor, modified by at least a dozen other hormones, enzymes, and food metabolites, control the cell proliferation.

Now it appears that the spleen can make the same kind of comeback, at least in children. Adults who must have their spleens removed rarely miss them, but children become more susceptible to meningitis. A few years ago, medical researchers noted that children whose excised spleens had been damaged by accident were less likely to get meningitis than those whose spleens had been removed because of disease. Howard Pearson and his colleagues at the Yale University School of Medicine found that, when ruptured, the delicate spleen left bits of itself scattered in the abdomen, which grew and gradually resumed the organ's obscure blood-cleansing functions. Now when they remove a spleen, many surgeons wipe it on the peritoneum (the tough membrane that lines the abdominal cavity) to sow replacement seeds.

Another late discovery of regeneration was made in the late 1950s, when several scientists learned that tadpoles, larval salamanders, and sometimes adult salamanders could restore up to about four inches of their intestines. Moreover, all adult amphibians could reconnect the cut ends even if they couldn't replace a missing section. Allan Dumont, one of my best friends during medical school and now Jules Leonard Whitehill Professor of Surgery at NYU-Bellevue, decided to check this potential in mammals after I told him about my work on rat limbs. He wanted to find out whether regeneration could be stimulated in mammals to solve one of a general surgeon's most vexing problems—poor healing of sutured ends of gut after a cancerous or degenerated segment has been cut out. Even a small opening can spill feces into the abdominal cavity, with disastrous peritonitis the result.

Like any good scientist, Al started from the basics. After several years he'd confirmed the earlier reports. When he cut pieces from the gut of adult frogs and newts and merely put the ends close together in the abdomen, 40 percent of the animals survived by quickly reconnecting the two ends and completely healing them in about a month, although even the newts didn't replace much of the lost length in his experiments. Gut regeneration actually involves several tissues; Al's cell studies showed a blastema quickly forming at the junction and then differentiating into smooth muscle, mucus cells, and the structural cells of the villi.

Naturally, when I was organizing a conference on regeneration in 1979, I invited Al to present his results. About a month before the meeting, after the program had already gone to the printer, he wanted to change the title of his paper, for he'd just finished some surprising work. He asked me, "What would you expect to happen if I took some adult rats, cut out a centimeter of gut, and dropped the two loose ends back into the abdomen?" Like any first-year med student, I said they'd be dead of peritonitis in two or three days. Well, 20 percent of Al's rats had reconnected their bowels better than surgery could have done, and were alive and healthy. When Al had given one group of animals a temporary colostomy above the experimental cut, the survival rate jumped threefold. The perfect healing of the test animals compared with the controls indicated that sutures actually interfered with regeneration producing unnecessary scars and adhesions.

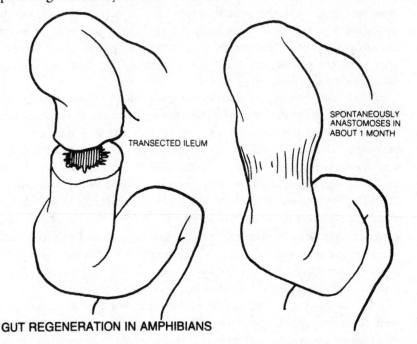

GUT REGENERATION IN AMPHIBIANS

No one knows for sure how the two cut ends find each other, but there's certainly some active search going on, for peritonitis sets in too quickly for the results to be due to chance. The process resembles a regrowing nerve fiber's search for its severed part, which may be conducted by electrical factors, a chemical recognition system, or both. Electrical potentials probably play the most important part, for recent research has found DC potentials at injuries on the peritoneum, and experimental changes in the peritoneum's normal bioelectric pattern attract the inner membrane enclosing the bowels, causing it to adhere to the site of the disturbance. Al has recently learned that, if the ends don't have to look for each other but instead are connected by a piece of Silastic tubing, rats can, like tadpoles, replace up to 3 centimeters of missing intestine. There's no reason to believe this technique couldn't be adapted to humans.

Even though we don't know enough yet to electrically stimulate intestinal healing, Al has proposed a preliminary test of regeneration in large mammals that could spare some patients a lifetime of misery. It's almost impossible to surgically rejoin the colon to the anus, and when sutures fail, the person ends up with a colostomy. Since the free end of the colon would be held near its proper position by the local anatomy, Al suggests replacing it without stitches and giving such animals a temporary colostomy upstream from the gap. If X rays later showed regrowth, the temporary colostomy would be closed, and the animal would have a continuous, healthy intestine. If even a few patients could be spared the indignity of living attached to a bag, the effort would be well worth the little—yet still nonexistent—funding required. Intensive study of the electrical details of gut healing would probably make surgery less devastating for many additional patients.

Exciting as the prospects in this survey are, they're by no means the only ones, or even the most spectacular, which are reserved for the following chapters. Many researchers are working to turn the breakthroughs of the last two decades to practical use. Even so, progress isn't nearly as fast as it could be, perhaps due to disbelief that such widespread self-repair is really possible for us. It's not only possible, it's nearly certain, given even a modest monetary push, for the "useful dispositions" foreseen by Spallanzani are within our reach.

Ten

The Lazarus Heart

Like Columbus, scientists sometimes stumble upon new continents when merely seeking a quicker trade route. Our research group had this good fortune in 1973.

We'd gone back to basics after learning how to dedifferentiate frog red blood cells and start regeneration in rat limbs. We decided to study nucleated red cells in a variety of creatures, hoping for leads toward better regrowth in mammals. Although their circulating erythrocytes have no nuclei, even mammals have young red cells, with nuclei, forming in the bone marrow. After severe bleeding, up to a fifth of those in the bloodstream may be immature nucleated types, as the marrow rushes them into service to make up for the loss. We surveyed the effect of direct current on red cells from fish, amphibians, reptiles, and birds. All of the cells responded, but in a different way for each species. We decided to have a more detailed look at the largest and hence most easily studied red blood cells available, those of our old friend *Triturus viridescens*, the common green newt.

A newt is so small that you can't just poke a needle into one of its veins and take a blood sample. The only practical way to get pure blood is to anesthetize the animal, slice open its chest, cut its heart in two, extract the blood with a pipette, and throw away the carcass.

As the phrase goes, we "harvested" blood from three newts each week by this method. One day, when Sharon Chapin had finished the chore, she asked me, "What would happen if I sewed these animals up?" I answered that, because their hearts had been destroyed, they would die within minutes, with or without sutures, from lack of oxygen to the brain, just as all our other amphibian blood donors had. We looked it up just to make sure. The standard works on regeneration all agreed that no animal's heart could repair major wounds. Unlike skeletal muscle, the cardiac variety had no satellite cells to serve as precursors for mature heart-muscle cells. In any case, the textbooks stated, the animal would die long before such repair could occur.

Next week Sharon put our three intended sacrifices in a bowl of water and with a straight face asked me if they looked healthy enough to use. I told her they looked fine. "Good!" she exclaimed. "These are the same three we used last week." Score one for the open mind!

Flabbergasted, I helped anesthetize and dissect this trio of miracles. Their hearts were perfectly normal, with no evidence of ever having been damaged in any way.

The Five-Alarm Blastema

Abruptly I changed my research plans. I asked Sharon to test a series of newts by cutting away large sections of their hearts and sewing up their chests, then killing some of the survivors every day and slicing, mounting, and staining the hearts for study under the microscope. Over 90 percent lived through the first operation, and several weeks later we had hundreds of slides ready for my examination and diagnosis. Unfortunately they all looked the same! Even those from *the day after* that horrendous mutilation showed only normal tissues with no sign of injury.

By now we knew we had come upon a first-class mystery and had better jettison our preconceptions. We reasoned that we could tell when regeneration was finished by finding out when the blood began flowing again. Under the microscope we could easily see blood cells streaming through capillaries in the transparent tail fins of lightly anesthetized newts. The motion stopped when we cut the heart, and restarted about *four hours* later. We sectioned a new series of hearts, this time covering the first six hours at intervals gradually increasing from fifteen minutes to one hour.

While waiting for the specimens, we rummaged more thoroughly through the literature for other reports on heart regeneration. There was evidence for very limited repair—but no true regeneration—of small heart wounds in a few animals. The process seemed limited to the very young. Even then, the results were of poor quality, combining a lot of scar tissue with only a little proliferation of nearby heart cells, but the mitotic component could be enhanced by various experimental aids.

In 1971, John O. and Jean C. Oberpriller, anatomists at the University of North Dakota School of Medicine in Grand Forks, reported that small wounds in salamander hearts healed this way but required two months. A year after that, the English edition of a book by Lev Polezhaev summarized several decades of Russian research, mainly on the hearts of frogs and lizards. Pavel Rumyantsev, now at the Cytology Institute of the USSR Academy of Sciences in Leningrad, had found in 1954 that newborn mammals (rats and kittens) could repair tiny puncture wounds, and recently he has proved the same capacity in the atria, or receiving chambers, of *adult* rat hearts. We even found a German report of 1914 claiming that human babies had sometimes regenerated small areas of their hearts damaged by diphtheria.

The Russians claimed some progress in extending this marginal native healing. In the late 1950s, N. P. Sinitsyn had repaired large holes (up to 16 square centimeters) in the hearts of dogs by covering the wounds with patches made of muscle sheath, canvas, suede, or other materials. Scar tissue still covered the outside, but the patch guided a thin layer of new muscle fibers forming along its inner surface. Using dogs whose wounds had already closed with scar tissue, Polezhaev then found he could induce heart muscle to fill in part of the gap by cutting away the scar and irritating the edges of the remaining cardiac muscle. Other Soviet researchers enhanced the muscle cell proliferation a little more with vitamins B_1, B_6, and B_{12}, various drugs, extra RNA and DNA, and heart tissue extracts or minces.

Despite such goads, heart regrowth was limited to very small injuries or the border zone around larger ones, and it always took several weeks. No one had even imagined that half a heart could restore its other half, much less in a matter of hours. I could hardly wait until the next batch of slides was ready.

They showed us an unprecedented type of regeneration. Where the missing part of the heart had been, a blastema formed in about two and a half hours. We saw no evidence of dedifferentiation or mitosis in the remaining heart-muscle cells, and indeed it would have been impossible for the processes we'd already studied to make a blastema in such a short time. Instead, the mass of primitive cells arose dramatically from the blood.

As soon as the salamander heart is cut open, blood pools around the wound and clots quickly, usually in about one minute, sealing the hole like wet plaster. Almost immediately, the nearest red blood cells crack open like eggs. Their nuclei, surrounded by a thin coating of cytoplasm, glide by some means yet unknown directly to the raw, frayed edge of the heart muscle and insinuate themselves into the tangle of dying and in-jured cells. To a biologist this sight is bizarre, uncanny. It's as though the engine of a passing car could walk up to a stranded truck, climb under the hood, and drive it away.

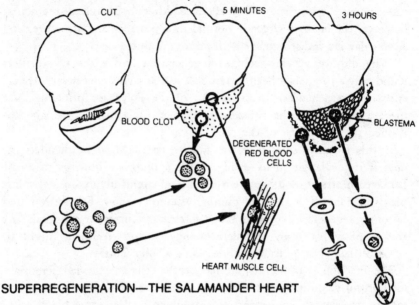

SUPERREGENERATION—THE SALAMANDER HEART

Farther away from the wound surface, the red cells also spill out their nuclei, but these cell yolks clump together, fusing their remaining cytoplasm to form a syncytium. Still farther away from the center of action, the red cells undergo the more leisurely dedifferentiation we ob-served in our frog fractures and DC culture studies. They turn into primitive ameboid cells that move toward the area of damage and attach themselves by pseudopods to the injured muscle fibers. In all of biology there's no precedent for these virtuosic cellular metamorphoses. In fact, they're so strange that most researchers have simply refused to believe in their existence or try the experiment for themselves.

All these changes are well under way within fifteen minutes. Soon afterward the extruded nuclei, the interconnected syncytial nuclei, and the ameboid cells are all dividing as fast as they can, building up the blastema. It's fully formed within three hours after the injury. By then its cells have already started to redifferentiate into new heart-muscle cells, synthesizing their orderly arrays of contractile fibers and con-necting up with the intact tissue. If the clot contained more blood cells than were needed, the extras outside of the area now degenerate, appar-ently so as not to get in the way of the repair work.

Meanwhile, the newt has survived by absorbing dissolved oxygen from the water through its skin. Now, at about the four-hour mark, there are enough new muscle cells to withstand contraction, and the heart begins pumping again, slowly. After five or six hours, most of the blastema cells have redifferentiated into muscle, which is still somewhat "lacy" or delicate compared with the established tissue. After about eight or ten hours, however, the heart is virtually normal in appearance and structure, and after a day it's indistinguishable from an uninjured one.

Why did we see this colossal regeneration, while the Oberprillers found only a tiny, slow healing response in the salamander heart? Apparently this was another manifestation of the Polezhaev principle. We made a big wound; they made a small one. Only massive damage unleashed the full power of the cells.

Is this fantastic cellular power forever restricted to salamanders, or does it reside latent in us, ready at the appropriate impetus to repair damaged hearts without problem-filled (and frightfully expensive) transplants of donated or artificial pumps? We don't know, but we've found no other regenerative process that's forever off limits to mammals. At this point we can only speculate on how such a treatment might be accomplished, but at least the idea isn't wholly fantasy.

The first job is to identify human target cells able to dedifferentiate into primitive totipotent cells. Bone marrow cells or immature erythrocytes, the nearest equivalents to amphibian nucleated red blood cells, are one obvious candidate population, especially since they seem to be the crucial cells in rat limb regeneration and the inner part of fracture healing. Fibroblasts despecialized by electrically injected silver ions might be used. Another possibility is lymphocytes, one class of infection-fighting white blood cells. In our lab we've demonstrated that they, too, can dedifferentiate in response to appropriate electrical stimuli.

Since newt-type heart regeneration doesn't occur naturally in mammals, we would probably have to grow a large mass of the target cells in tissue culture. Then, with the patient on a heart-lung machine, the surgeon could cut away scar tissue and otherwise freshen the wound if it wasn't recent enough, then apply enough of these ready-made pre-blastema cells to fill the defect. They would be held in place by a blood clot, sutured pericardium, or some type of patch. Then, assuming we'd learned the electrical parameters already, electrodes would induce nuclear extrusion, dedifferentiation, consolidation with surrounding muscle, and the final transformation into normal cardiac muscle. The current would probably have to be adjusted throughout the process to get its various steps in synchrony, and vitamins or drugs might be used to enhance mitosis or protein synthesis. Once the scar had been removed, the instructions as to what cells were needed would come from the surrounding healthy heart muscle.

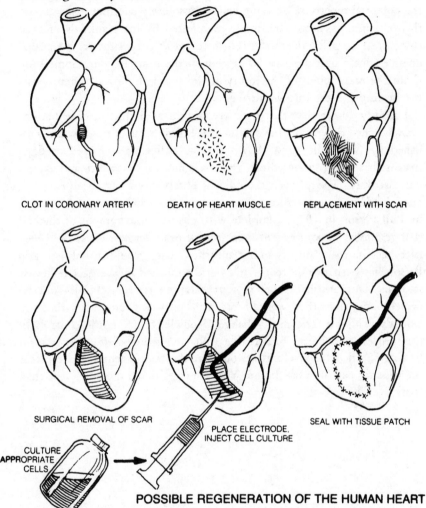

CLOT IN CORONARY ARTERY   DEATH OF HEART MUSCLE   REPLACEMENT WITH SCAR

SURGICAL REMOVAL OF SCAR   PLACE ELECTRODE. INJECT CELL CULTURE   SEAL WITH TISSUE PATCH

CULTURE APPROPRIATE CELLS

POSSIBLE REGENERATION OF THE HUMAN HEART

In the salamander this process takes about six hours. Since this exceeds the current limits of "machine time" for artificial circulation in humans, we would have to extend the capacities of heart-lung devices or else speed up the cellular processes. Obviously there's a long road of experiment to travel before we can be more specific about techniques. One of the things we must learn is whether the newt's electrocardiogram shuts down during repair. We must know how its presence or absence relates to the current of injury and other electrical factors in this novel method of blastema formation.

Personally, I'm sure we can get the human heart to mend itself. As a result of being confronted by this wonder in newts, I'm convinced that the potential repertoire of living cells is absolutely enormous, far greater than the healing powers normally manifested by most animals or even those dreamed of by doctors. Even in the newt this "superregeneration" doesn't appear unless 30 to 50 percent of the heart is gone. Something about the massiveness of the injury or the approach of death then boosts the healing process into overdrive.

I readily admit that the discovery sounds a bit like science fiction, even as toned down into the subdued technical prose of our report, published by *Nature* in 1974. I had trouble believing it myself at first. Because it seemed so incredible, there was no rush to confirm and extend our discovery. Today, even though our observations have been corroborated by University of Michigan anatomist Bruce Carlson in 1978 and by Phil Person in 1979, complete with electron micrographs of the cell changes, most biologists still don't accept heart restoration as fact. Perhaps because the reality is so outlandish, Carlson wouldn't publish, and Person has been unable to get his work published in the peer-reviewed journals. Our original paper of ten years ago is still officially unconfirmed, and the other workers are still puttering around with little wounds. This attitude must change. Knowledge about the controls of this process will be of incalculable value to medicine, for this is ideal healing. Spilled blood closes a wound at the body's center and replaces the missing part in a few hours. You can't get much more efficient than that.

Eleven

The Self-Mending Net

Spinal paralysis is the most devastating of injuries and also one of the commonest; it afflicts over half a million Americans, including fifteen thousand new sufferers every year. Until recently their outlook was absolutely bleak, for the human central nervous system (CNS) had no known regenerative capacity whatsoever. Only if part of the spinal cord remained unsevered was some recovery possible with physical therapy. Now, however, there's hope that we'll soon be able to coax nerve cells into reestablishing the proper connections across the damaged section and thus return the use of arms, legs, sexual and excretory organs, respiratory muscles, and the sense of touch to quadriplegics and paraplegics. In one way or another, this dream involves making human nerve cells behave more like those in simpler animals.

The neuron is the basic unit of all nervous systems. It consists of a cell body, containing the nucleus and metabolic organelles, surrounded by dozens of filaments that carry messages in and out. The incoming dendrites predominate in sensory neurons. There's usually only one motor fiber, or axon, which carries the neuron's outgoing messages to dendrites of other neurons or to receptors on muscle or gland cells. An axon, often several feet long, is the principal fiber of a motor neuron, which relays orders from the brain or spinal cord to the tissues and organs.

All neuron cell bodies reside in the brain and spinal cord. Only their axons and dendrites extend outward, forming the peripheral nerves that connect every part of the body with the CNS. Other fibers connect certain sensory and motor neurons within the spinal cord, creating reflex arcs, like those that jerk our hands from hot stoves without our having to send the impulse all the way to the brain for instructions. Still other fibers connect spinal neurons with those in the brain, and in the brain itself the interconnections reach such a density that each nerve cell may hook up with as many as twenty-five thousand others.

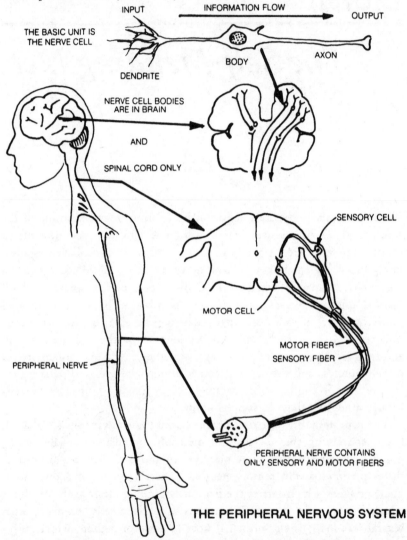

THE PERIPHERAL NERVOUS SYSTEM

Except for a few specialized components like the naked fiber tips that enter into neuroepidermal junctions, all parts of every neuron are swaddled in various types of perineural cells. In the brain there are several kinds, collectively called the glia, in which the neurons are embedded like hairy raisins in a pudding. The cell bodies in the cord also are surrounded by glial cells, but their axons and dendrites, which include the fibers of the peripheral nerves, are surrounded by Schwann cells. These form tubes, made up of spiraling layers of membrane rich in a fatty substance called myelin, around some of the largest fibers. A third type, ependymal cells, line the four cavities within the brain, or ventricles, and the narrow central canal that runs the length of the spinal cord. These cells are close relatives, having all developed from the same part of the ectoderm, or outer cell layer, that formed the primitive neural tube in the embryo. The nervous system actually consists of several times more perineural cells than neurons.

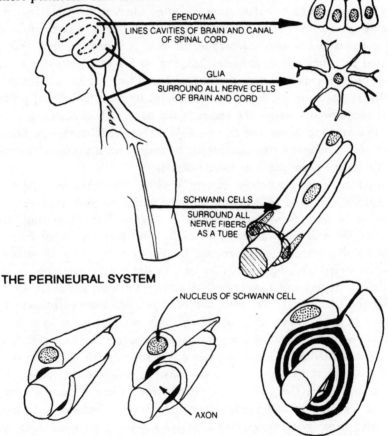

EPENDYMA
LINES CAVITIES OF BRAIN AND CANAL
OF SPINAL CORD

GLIA
SURROUND ALL NERVE CELLS
OF BRAIN AND CORD

SCHWANN CELLS
SURROUND ALL
NERVE FIBERS
AS A TUBE

THE PERINEURAL SYSTEM

NUCLEUS OF SCHWANN CELL

AXON

SCHWANN CELL SHEATH AROUND A PERIPHERAL NERVE FIBER

Until recently the perineural cells were considered merely a "packing tissue," whose only job was to insulate and support the neurons. We've now learned that they play a major role in getting nutrients to the neurons. They also help control the diffusion of ions through the nerve cell membrane and hence regulate the speed of impulse firing, even to the point of inhibiting seizures, the random spread of impulses in the brain. They may also have an important part in memory, and they probably conduct the direct currents so important to regeneration. They're essential to healing wherever it occurs in nerve tissue.

Peripheral Nerves

Peripheral nerve fibers can regrow—otherwise we'd lose sensation whenever we cut a finger—but neurons and their fibers in the CNS cannot. The peripheral nerve's cell body survives, safe in the cord or brain, and the cut end of the attached part of the fiber is sealed off. The outer, severed part dies and degenerates; some of its Schwann cells digest it, along with the now-useless myelinated membrane layers. The empty Schwann tube remains, however, and begins to grow toward the proximal fiber (the one nearer the center of the body), whose Schwann cells are also growing across the chasm. When these cells meet, the nerve fiber grows along its reconnected sheath and eventually makes contact with the same terminals it originally served.

In salamanders this process is very efficient. The Schwann cells can cross large gaps, and an experimenter who wishes to work with denervated limbs must be diligent to keep the nerves from reentering. In humans, the two ends of the tube usually can't find each other over a distance of more than a centimeter. In that case the proximal sheath with its intact nerve fiber hunts for its opposite number by growing in an ever increasing spiral, apparently searching for some signal from the distal end (the part farther from the body center). Since each nerve is formed of many fibers, these spiraling tubes entangle each other in a lump of nerve tissue called a neuroma. Neuromas are painfully sensitive and often must be cut away. Occasionally a surgeon can move the two ends of the nerve close enough for the Schwann cells to make contact. If that's impossible, the gap may be closed with a piece of nerve grafted from a less important peripheral nerve that can be sacrificed. Unfortunately, nerve grafts don't take reliably, and other methods, such as making artificial channels with tiny plastic tubules, are still in the experimental stage.

We don't know why the salamander's peripheral nerve regrowth is so much more effective than ours, but I surmise that its more efficient DC electrical system accounts for the difference. If the locator signal is electrical, it should be possible to augment it in humans so as to grow nerve fibers over longer distances. Beginning with a 1974 report from David

H. Wilson of the Leeds General Infirmary in England, there have been some interesting claims that pulsed electromagnetic fields have speeded recovery of limb function in rats after peripheral nerve damage, but the effect hasn't yet been substantiated for humans. If these findings hold up, we may soon be able to boost nerves past their 1-centimeter limit, even if the action is indirect, and a thorough investigation of the electrical basics could drive nerve regrowth to even greater lengths.

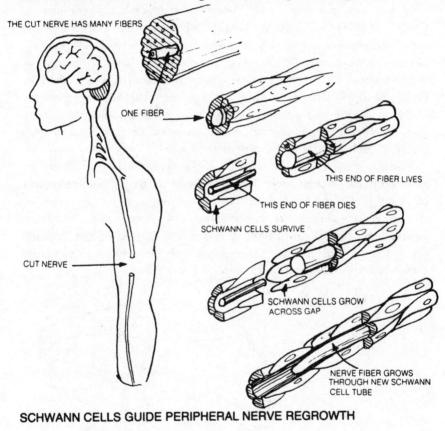

THE CUT NERVE HAS MANY FIBERS

ONE FIBER

THIS END OF FIBER LIVES

THIS END OF FIBER DIES

SCHWANN CELLS SURVIVE

CUT NERVE

SCHWANN CELLS GROW ACROSS GAP

NERVE FIBER GROWS THROUGH NEW SCHWANN CELL TUBE

SCHWANN CELLS GUIDE PERIPHERAL NERVE REGROWTH

The Spinal Cord

A sad and crucial difference separates peripheral fibers from those in the human spinal cord, for the latter don't reconnect over even a fraction of a centimeter. However, in most injuries relatively few of the neurons themselves are killed. It's important to realize that most of the cord cells below the injury don't die. The reflex arcs remain intact. In fact, reflexes are stronger than normal, because the neurons are now disconnected from the regulating influence of the brain. For the same reason, the broken bones of paraplegics heal in half the normal time, whereas a bone will heal very slowly or not at all if its *peripheral* nerve supply has been cut. Only the communication between brain and spine is silenced in paraplegia, and that makes all the difference.

Spinal fibers *do* reconnect in some animals, notably goldfish and, as you might expect, salamanders. Their ability seems to decline dramatically with age, however. Jerald Bernstein, a neurophysiologist now at George Washington University Medical School who has studied goldfish spinal regeneration extensively, has found that one-year-old fish heal almost all of the damage. This competence declines to about 70 percent at two years and 50 percent at three. Since salamanders aren't raised in biological supply houses but rather collected from the wild, any group is likely to include young and old individuals, making comparisons difficult. In our lab we found that cord regeneration isn't uniform in salamanders, probably due to age differences.

Maturity may reduce the response of the ependymal cells, which are responsible for the first step. They proliferate outward from the central canal and bridge the gap in a few days. Marc Singer, in a recent study of this process, concluded that the ependymal cells extend "arms" radiating outward, which line up like the spokes of wheels stacked one atop another, forming channels for the regrowing fibers to follow. The nerves then reestablish their continuity within a few weeks.

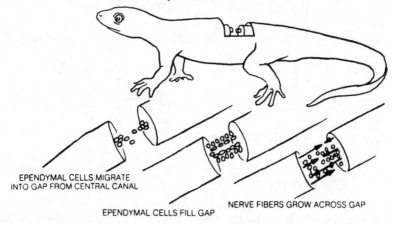

EPENDYMAL CELLS MIGRATE
INTO GAP FROM CENTRAL CANAL

EPENDYMAL CELLS FILL GAP

NERVE FIBERS GROW ACROSS GAP

SALAMANDER SPINAL-CORD REGENERATION

Bernstein also found that there's a critical period during which regrowth must be completed or it will fail. After cutting the cords of goldfish, he inserted Teflon spacers to block regeneration. The normal processes took place, but of course the cells couldn't penetrate the divider. After the cellular activity had died down, Bernstein removed the barriers, but there was no further change. However, when he then cut off each damaged end, producing an even larger gap and reinjuring the cord, the cells started from scratch and healed the defect completely. Thus there's good reason to believe that even long-standing spinal injuries can potentially be regenerated if we can extend the basic capabilities of human cells.

One would expect to see some healing response in mammals, even if it fell short. After all, we only need the elongation and reattachment of fibers, which does take place in peripheral nerves. Instead the opposite happens. The cord cells die a short distance above and below the injury. Cysts form near the ends, and, instead of ependyma, scar tissue fills the gap. Only after this destruction is there an abortive attempt at regrowth. In humans this amounts to only a few millimeters of fiber elongation many months after the injury. By then it's too late; the ependymal cells and nerve fibers can't penetrate the scar.

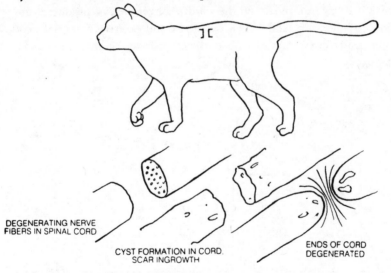

DEGENERATING NERVE
FIBERS IN SPINAL CORD

CYST FORMATION IN CORD.
SCAR INGROWTH

ENDS OF CORD
DEGENERATED

CYSTS AND SCARS PREVENT CORD REGROWTH IN MAMMALS

Why the difference between salamanders and mammals? The reason may lie in the cord's immediate response. In all animals the injury instantly results in spinal shock, during which all neuronal activity is profoundly depressed, especially in the part of the cord still connected to the brain. Even the simplest reflexes disappear. As the shock wears off, the cord *below* the injury becomes hyperactive. Its reflexes become tremendously exaggerated and lead to spastic paralysis of the muscles. The interesting difference is in the duration of shock. In young salamanders and goldfish it lasts only a few minutes, but it may endure for over a week in old ones. In mammals it takes even longer to wear off—as long as six months in humans.

We made some electrical measurements on salamander and frog spines in our lab. The injured area turned out to be strongly positive during spinal shock, even though all direct-current flow ceased in the entire cord and in the peripheral nerves arising from the part below the trauma. Then, as the shock resolved, a steadily increasing negative potential appeared, its size reflecting the amount of outgrowth by epen-. dyma and nerve fibers. We found that we'd only *re*discovered these potentials, however. G. N. Sorokhtin and Y. B. Temper had made the same measurements at the Khabarovsk Medical Institute twenty years before. The patterns of shock and polarity both correlated, not only with the cell activity, but with the end result of regenerative success or failure. A few minutes of shock and a correspondingly short period of positivity led to full repair of the cord. Longer delays produced incomplete regeneration, and, when the shock and positive potential persisted for five to eight days or longer, the salamander became completely paraplegic.

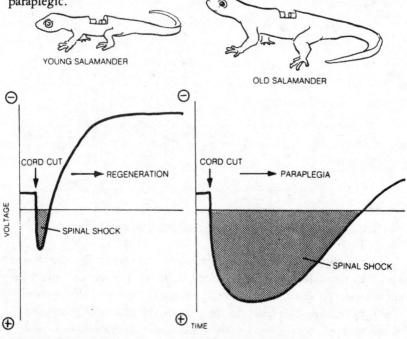

YOUNG SALAMANDER

OLD SALAMANDER

CORD CUT → REGENERATION

CORD CUT → PARAPLEGIA

SPINAL SHOCK

SPINAL SHOCK

VOLTAGE

TIME

SPINAL SHOCK AND AGE INHIBIT CORD REGROWTH

As far as I know, the only electrical measurements of spinal shock in mammals were preliminary ones done at our laboratory in conjunction with Carl Kao of the VA hospital in Washington, D.C. We tested the severed cord ends in cats for twenty-four hours and found only an increasing positive potential. The situation seemed quite similar to the electrical difference between salamander and frog limbs. As in most instances, positive potentials appeared to inhibit constructive cellular activity while negative ones fostered it.

An experiment Kao did several years ago provided some supporting evidence. Kao made two cuts through the spinal cord in each of several cats, producing a central fragment about 5 millimeters long, separated from each end. He then grafted pieces of sciatic nerve as spacers in the two cuts. Typical degeneration with cysts occurred in each end of the cord but not in the isolated piece. In fact, this part showed some growth of its ependyma and nerve fibers. The small piece was probably isolated from the positive potentials produced in the rest of the cord. Hence it escaped inhibition and grew. It seems the prolonged electrical positivity of spinal shock is the main roadblock in the way of human cord repair.

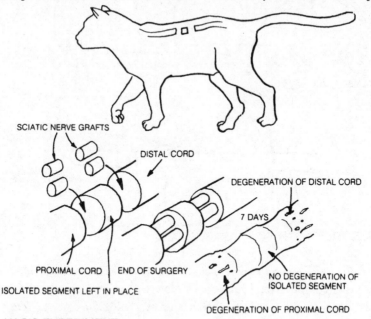

SCIATIC NERVE GRAFTS

DISTAL CORD

DEGENERATION OF DISTAL CORD

7 DAYS

PROXIMAL CORD | END OF SURGERY

ISOLATED SEGMENT LEFT IN PLACE

NO DEGENERATION OF ISOLATED SEGMENT

DEGENERATION OF PROXIMAL CORD

KAO'S EXPERIMENT

It should be possible to cancel that polarity and replace it with a growth-stimulating negative one, using a properly shaped electrode. Older injuries in which spinal shock has subsided might require a different input of current, as well as surgical removal of scar and cysts. The electrode material would have to be chosen carefully, for some metals are toxic to nerve cells. Also, humans have a low ratio of ependyma compared to spinal neurons, so we might have to add more. However, it should be relatively easy to culture more ependymal cells from a sample of the patient's own, and then inject them when we put in the electrode.

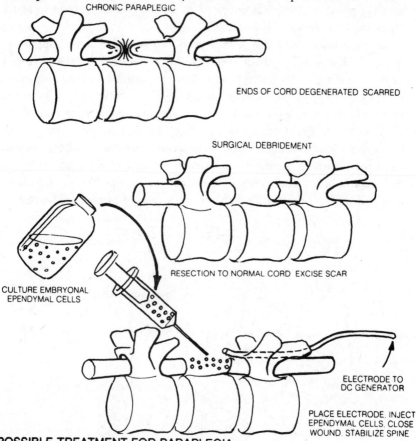

CHRONIC PARAPLEGIC

ENDS OF CORD DEGENERATED SCARRED

SURGICAL DEBRIDEMENT

CULTURE EMBRYONAL EPENDYMAL CELLS

RESECTION TO NORMAL CORD EXCISE SCAR

ELECTRODE TO DC GENERATOR

PLACE ELECTRODE. INJECT EPENDYMAL CELLS. CLOSE WOUND. STABILIZE SPINE

POSSIBLE TREATMENT FOR PARAPLEGIA

Not too many years ago spinal accident victims usually died of infections or other complications quite soon. Now we can prolong their lives, but only at enormous social, financial, and psychological cost. Looking ahead, as in the case of heart damage, we now have hope for releasing regeneration in humans. Actually, the outlook for spinal regrowth is more promising: The cellular processes are more familiar, and there are a few groups, like the American Paralysis Association, that sponsor research more imaginatively than the government agencies. Thus the electrical problems in spinal healing may be tackled sooner than in other fields.

The public imagination has been captured by the computerized muscle-stimulation techniques being developed by Jerrold Petrofsky, an engineer at Wright State University in Dayton. The nationally televised sight of his patient Nan Davis and other paraplegics taking tentative steps and pedaling tricycles with their own muscle power was tremendously exciting. But if we can get the body to do the same things by itself, that will be even better. Any amount of regeneration would only make other techniques more effective. Even restoring 10 percent of lost function would be an unimaginable blessing to those who are now helpless. I feel the electrical manipulation of spinal shock must be tested vigorously now, for this is perhaps the one area where the barriers of tragedy are closest to being broken.

The Brain

It might seem foolish to expect any regeneration in the most complex of all biological structures, the brain, yet salamanders, some fish, and most frogs in the tadpole stage can replace large parts of it, including the optic lobes and the olfactory lobes, or forebrain, the part from which our prized cerebral hemispheres developed in the course of evolution. Replacement depends on ingrowth of remaining sensory nerves, the olfactory nerves in the case of the forebrain and the optic nerves for the optic lobes. When these nerves grow back into the area where brain has been destroyed, they stimulate the ependymal cells in the brain ventricles, which proliferate outward into the damaged part and then differentiate into new neurons and glial cells. If the animal's nose or eyes are removed so that the injury zone receives no nerve input, no regeneration occurs.

Thus brain regrowth begins much like that of limbs, with the connection of nerve fibers to an epithelial tissue. The ependyma, remember, is embryologically a close relative of the epidermis, and in fact can be considered the central nervous system's "inner skin." Since the electrical environment produced by the neuroepidermal junction is what stimulates cells to dedifferentiate and divide in the salamander limb stump, and since we started limb regeneration in the rat by crudely mimicking this signal, it seems likely that a similar stratagem could induce brain regeneration in animals normally lacking this ability.

A form of shock, called the spreading depression of Leão after its discoverer, neurologist A. A. P. Leão, occurs after brain injuries. Starting at the site of damage, it extends in all directions until the entire cortex becomes electrically positive and all its neurons shut down. Leão studied it only in response to small injuries, when it persists for a few hours. We don't know whether it occurs in the salamander or how long it lasts after major damage to the mammalian brain. Concerted study of Leão's depression combined with experiments in electrically stimulating the ependymal cells could open the way to self-repair of the human brain.

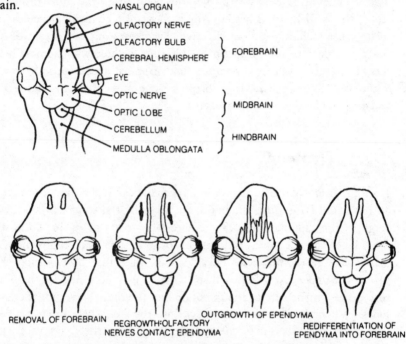

AMPHIBIANS CAN REGENERATE LARGE PARTS OF THE BRAIN

Recovery from stroke and head wounds taught us long ago that the brain has a great deal of plasticity; that is, it can reorganize so that undamaged regions take over tasks formerly done by the lost cells. Supplementation of this ability with even a small amount of regeneration might make recovery nearly complete for many brain-damaged people. For the first time in history, neurologists can hope to progress from describing the brain and cord to mending them. As Geoffrey Raisman of London's Laboratory of Neurobiology recently reminded his colleagues: ". . . no immutable natural laws have been discovered that forever rule out repair of the nervous system."

Twelve

Righting a Wrong Turn

Good and evil often sprout from the same tree, in the body as in Eden. Nothing illustrates this paradox better than cancer. Today, because of breakthroughs in genetics, thousands of scientists are searching for oncogenes, bits of DNA that are presumed to pull the trigger that fires the malignant bullet. It has been known for a long time, of course, that cancer isn't inherited through egg and sperm the way hemophilia is. However, many have postulated that the immediate cause of cancer may be genetic changes in somatic cells. Normally suppressed genes held in an unnoticed corner of our genetic bookshelves since long ago in our evolution might be dusted off only when other bodily conditions are "just wrong." While the premise of this idea is apparently true, biologists have recently concluded that the difference between a normal gene producing a normal protein and one that could theoretically cause cancer is a single "typographical error" in a whole chapter of amino acid sequences. Such mistakes happen so often that we would all be riddled with cancer from infancy if that were all it took to start the disease. Something else must go awry before a few misspellings can turn the whole library into gibberish.

Three basic criteria by which a doctor diagnoses cancer must serve as the starting point in solving the mystery of its cause. First of all, the disease always arises not from an alien germ but from a formerly normal cell of the host's body, and the cancer cells are more primitive than their healthy precursors. Moreover, this atavism reflects the seriousness of the disease: The simpler the cells, the faster they grow and the harder they are to treat, whereas a tumor that still resembles its tissue of origin is less malignant.

The second criterion is growth rate. Cancer cells multiply wildly, in contrast to the slow, carefully controlled mitosis of normal cells. Going hand in hand with this uncontrolled proliferation is a similar lack of control in the structural arrangement of the cells. Their membranes don't line up in the normal, specific ways, and they form a jumbled mass instead of useful architecture. As a further result of runaway multiplication, cancer doesn't observe the "boundary laws" of normal tissue. Instead it encroaches imperialistically upon its neighbors. In addition, since the cells don't adhere in any kind of structure, some of them are constantly breaking off, flowing through the blood and lymph, and setting up colonies—metastases—throughout the body.

The third basic criterion of cancer is metabolic priority. The diseased tissue greedily takes first choice of all nutrients circulating in the blood; the healthy part of the body gets what's left over. As the tumors disseminate and grow, they consume all available food, and the host wastes away and dies.

We can make one crucial observation at this point: Except for the lack of control, all three characteristics—cell simplicity, mitotic speed, and metabolic priority—are hallmarks of two normal conditions, embryonic growth and regeneration.

When considering the similarities between an embryo and a tumor, it's important to keep in mind one difference. Even though contained within the body of its mother, the embryo is a complete organism, and the controls over its cells are primarily its own, not those of an adult. Over thirty years ago in Switzerland, G. Andres probed this relationship by implanting frog embryos in various body tissues of adult frogs. Whenever the host didn't simply reject the graft, the embryo degenerated into a highly malignant metastasizing tumor. As a result, Andres proposed a theory of cancer that remains provocative today: A normal cell becomes cancerous by dedifferentiation. This change is not dangerous per se, according to Andres, but, because it occurs in a postfetal animal, the controls that would normally hold these neo-embryonic cells in check aren't working.

Cancer's relationship to regeneration is even more interesting. In the latter, a rapid growth of primitive cells having metabolic priority occurs in an adult, but with proper control as in an embryo. Those animals that regenerate best are least susceptible to cancer. In general, as complexity increases up the evolutionary ladder from worms to humans, regeneration decreases and cancer becomes more common. Although salamanders stand about midway in degree of complexity, they're perhaps the least specialized of all land vertebrates. They have tremendous regenerative abilities and almost no cancer. Even to give them tumors in the laboratory requires much effort. Adult frogs, on the other hand, have bodies that are much more specialized for their amphibious way of life; they regenerate very little and are subject to several kinds of cancer.

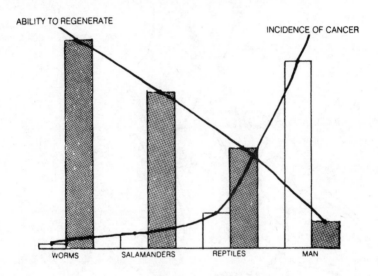

AS REGENERATION DECREASES, CANCER INCREASES

In 1948 Meryl Rose decided to see whether the environment of a salamander's regenerating limb could control the primitive cells of cancer as well as those of the blastema. He took pieces of a type of kidney tumor common in frogs and transplanted them to the limbs of salamanders. These tumors took better than most, and soon killed the animals when allowed to spread unchecked. However, when Rose amputated the leg just below or through the malignancy, normal regeneration followed, and the cancer cells dedifferentiated more fully as the blastema formed. Then as the new leg grew, the former frog tumor cells redifferentiated along with the blastema. The frog cells were easily distinguished from salamander cells by their smaller nuclei, and microscopic study showed frog muscle mixed in with salamander muscle, frog cartilage cells amid salamander cartilage, and so on.

This monumentally important experiment proved Rose's hypothesis that regeneration's guidance system could control cancer, too. It implied that cancer cells weren't special but merely embryonic cells in a post-embryonic body. Rose's work led directly to Andres's theory a few years later.

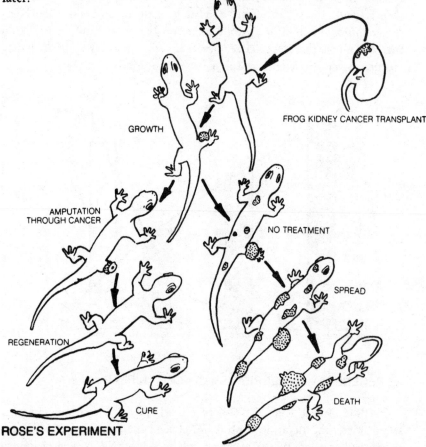

ROSE'S EXPERIMENT

Unfortunately, biology was still firmly gripped by anti-dedifferentiationism, and these ideas were partly ridiculed, partly ignored. The reaction held back cancer research for decades, because the dogma implied that carcinogenesis, like differentiation, was irreversible—once a cancer cell, always a cancer cell. As long as this view was sacred, the only possible way to cure cancer was to cut it out or kill it with drugs and X rays. We've been beating that dead horse for fifty years now with tragically modest increases in survival rates. Surgery works only against tumors that haven't yet spread. Chemotherapy depends on differences between malignant and healthy cells. However, the differences aren't as great as we would like, because cancer arises, not in some distant swamp, but from a slight change in our own cells. Therefore chemotherapy and X rays inevitably produce some damage in normal cells, too. Doctors, being only human and sharing in their patients' pain and terror, have devised ever more ferocious treatments. In our war on cancer we've stampeded ourselves into a sort of Vietnam syndrome: To destroy our foes, we're killing our friends. As Walt Kelly's cartoon character Pogo observed in that context, "We have met the enemy and he is us."

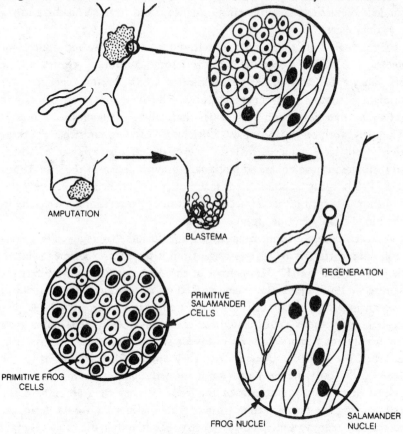

AMPUTATION

BLASTEMA

PRIMITIVE
SALAMANDER
CELLS

REGENERATION

PRIMITIVE FROG
CELLS

FROG NUCLEI

SALAMANDER
NUCLEI

CANCER CELLS RETURNED TO NORMAL BY REGENERATION

A Reintegrative Approach

But surely, you may be thinking, if this theory of cancer had any potential for a cure, the research establishment would have considered it. And surely there would be some supporting evidence. Unfortunately, even though the detooling and retooling of cells have now been accepted by all of biology, the old habits still persist throughout most of the granting hierarchy. A few years ago, for example, I met a young research fellow at the National Cancer Institute who wanted to study the regeneration-cancer link. He even showed me his proposal, an excellent one. I told him he was asking for trouble if he submitted it to NCI, but he said his boss approved and he was sure he'd get the grant. A month later he was forced out of the institute, and the project has never been funded. Nevertheless, supporting evidence, from research on the periphery, does exist.

Since regeneration can't occur without the stimulus and control of nerves, one would expect them to exert some controlling effect on cancer. They apparently do. As far back as the 1920s, several experimenters implanted tumors into denervated areas. Without exception the cancer cells took root better and grew faster than where the nerves were intact. The early work on this point was criticized on the grounds that denervation might have reduced the efficiency of the circulatory system, which in turn would have enhanced malignant growth. Then in the mid-1950s and 1960s more sophisticated techniques established the same relationship. Absence of nerves accelerated tumor growth, and variations in the blood supply had no significant effect.

Further evidence confirming Rose's conclusion that regenerative controls caused tumors to regress came from a series of experiments by F. Seilern-Aspang and K. Kratochwil of the Austrian Cancer Research Institute in 1962 and 1963. They worked on salamanders, but instead of implanting frog tumor cells they induced skin cancer with large, repeated applications of carcinogenic chemicals. With persistence they eventually got tumors that would invade subsurface tissues, metastasize, and kill the animals. In one series they applied the carcinogen to the base of the tail; the primary tumor formed there, and metastases appeared at random in the rest of the body. If they then amputated the tail, leaving the primary tumor intact, this malignancy would disappear as the tail regrew. Cell studies showed that it didn't die or degenerate but apparently reverted to normal skin. Furthermore, all the secondary tumors vanished, too, as though they were being operated by remote control from the main one. The salamander ended up with a new tail and no cancer. However, if the primary tumor was at a distant point on the body, amputation of the tail had no effect. Even though the tail regenerated, the main cancer and its offshoots all progressed, and the animal died.

This research, combined with Rose's, indicates that regeneration near a primary tumor can make it regress by reverting to its normal tissue type. I doubt that there's anything special about legs or tails; I would predict that regrowth in any part of the body, as long as it was near the primary tumor, would have the same effect. The key to regression appears to be a change in the malignancy's immediate neighborhood. The electrical currents in nerve and particularly in the neuroepidermal junction seem likely candidates, since they suffice to start regeneration in animals normally incapable of it.

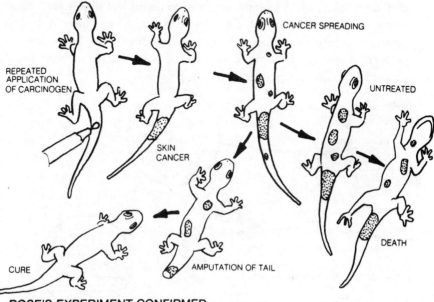

ROSE'S EXPERIMENT CONFIRMED

There's abundant evidence that the state of the *entire* nervous system can affect cancer. Back in 1927 Elida Evans, a student of Carl Jung, documented a link between depression and cancer in a study almost totally neglected in the intervening years. In a long-term project begun in 1946 by Dr. Caroline Bedell Thomas at Johns Hopkins School of Medicine, students were given personality tests, and the occurrence of disease among them was charted over several decades. In this and later studies, a high risk of developing cancer has been correlated with a specific psychological profile that includes a poor relationship with parents, self-pity, self-deprecation, passivity, a compulsive need to please, and above all an inability to rise from depression after some traumatic event such as the death of a loved one or loss of a job. In such a person, cancer typically follows the loss in a year or two.

Several physicians have found they can greatly increase cancer patients' chances of a cure with biofeedback, meditation, hypnosis, or visualization techniques. Several years ago O. Carl Simonton, an oncologist, and

Stephanie Matthews-Simonton, a psychologist, began using all these methods, with emphasis on having patients develop a clear picture of the cancer and their body's response to it. For example, a patient might spend a meditative period each day imagining white blood cells as knights on white horses defeating an army of black-caped marauders. When the Simontons tabulated their first results in "terminal" cases, they found that, of 159 people expected to die in less than a year, those who eventually did succumb lived twice as long. The cancer had completely regressed in 22 percent and was receding in an additional 19 percent. These results have held up, and visualization is now being adopted in some other cancer-treatment programs.

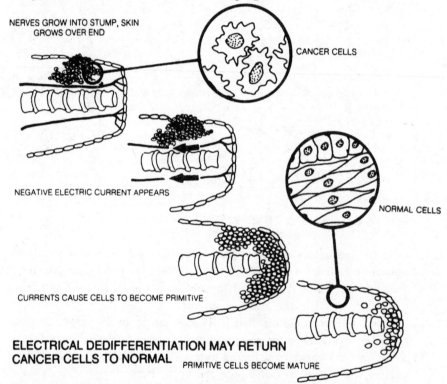

NERVES GROW INTO STUMP, SKIN GROWS OVER END

CANCER CELLS

NEGATIVE ELECTRIC CURRENT APPEARS

NORMAL CELLS

CURRENTS CAUSE CELLS TO BECOME PRIMITIVE

ELECTRICAL DEDIFFERENTIATION MAY RETURN CANCER CELLS TO NORMAL PRIMITIVE CELLS BECOME MATURE

Penn State psychologist Howard Hall, testing hypnosis for boosting white blood cell activity, found a 40 percent increase in cell counts among his younger, more responsive subjects just one week after the trance session. New Haven surgeon Bernie Siegel has developed the Simontons' methods further with therapy groups called Exceptional Cancer Patients. Having patients make drawings to reveal their true psychological state sans defenses, then working toward a comprehensive change in outlook (total CNS response) to mobilize the will to live, Siegel has helped his patients greatly improve the quality and quantity of their lives, as compared to clinical prognoses. This approach also enhances chemotherapy's effectiveness while minimizing its side effects, and it dramatically increases the likelihood of a "miracle"—total regression and cure of the cancer.

All this really shouldn't be so surprising. Under hypnosis the mind can completely block pain, and research described in the next chapter has shown that it does so by changing electrical potentials in the body. How can we be sure it couldn't create appropriate electrical changes around a tumor and melt it away? There are still major problems with these psychological approaches, however. Only a minority of people are able or willing to muster the high level of dedication needed to make them work, even under the gun of death. Moreover, they require time, the very thing a cancer patient is short of, and they often don't produce a complete cure even when practiced diligently. Still, they're encouraging signs that the tide is beginning to turn from the warfare mode to the simpler—more elegant, as a mathematician would say—ideal of changing the cellular environment that allows a tumor to flourish.

Could we really do the job more directly by applying the proper electrical input to aim a surefire regenerative influence at the tumor? I'm sad to say that most of the few researchers who've tried electricity against cancer have used the "kill the enemy" approach. Tumors are somewhat more sensitive to heat than normal tissue, so some doctors are using directed beams of microwaves to cook them without, it's hoped, destroying too many healthy cells. FDA approval for general use of this method is expected soon. It has been known since the time of Burr and Lund that growing tissue is electrically negative, and cancer is the most negative of all. Hence some researchers have tried to inhibit tumor growth by canceling the offending polarity with positive current. Early reports were encouraging, but we now know that toxic metallic ions are released from most positive electrodes, so this method must be tested with great caution.

Only one research team has sought a reintegrative effect of electric current. In the late 1950s Carroll E. Humphrey and E. H. Seal of the Applied Physics Laboratory at Johns Hopkins tried pulsed direct currents on standardized fast-growing skin tumors in mice. Even though they used both positive and negative polarities, their results seemed sensational. In one series they got total remission in 60 percent of the test animals after only three weeks; all the control mice had died by then. In another series the control tumors averaged seven times the size of the ones treated with current. Unfortunately, present evidence doesn't really support this approach either. In my group's experiments with human fibrosarcoma cells *in vitro,* negative and positive currents *both* speeded up growth by over 300 percent. On the other hand, as mentioned in Chapter 8, we found that we could suspend mitosis in the fibrosarcoma cells with silver ions injected by minute levels of positive current. During one day of exposure, the cells appeared to dedifferentiate completely, and they stopped dividing for a month without additional treatment, even though we changed the nutrient medium regularly. Obviously, this entire subject needs to be investigated more thoroughly.

Some researchers believe pulsed electromagnetic fields may be of some benefit in treating cancer. Art Pilla, working with Larry Norton and Laurie Tansman of New York's Mount Sinai School of Medicine, as well as William Riegelson of the Medical College of Virginia, claims to have found a pulse sequence that significantly increases the survival time of mice with cancer. So far, these experimenters say they've increased the effectiveness of chemotherapy in lab animals with PEMF, but haven't found a pulse pattern that consistently regresses tumors *in vivo,* although Pilla and Steve Smith have been able to transform malignant lymphoma (lymph node cancer) cells into benign fibroblasts in culture.

The claim that PEMF may retard cancer in animals is seriously flawed, however. In these experiments the *entire* animal was exposed to the field, not just the part with the cancer. The pulsing field (like almost any time-varying magnetic field) induces a stress response in the animal (see Chapter 15). For a short while this increases the activity of the immune system, which slows the growth of the tumor. However, the field's effect on the tumor itself is to speed it up, and, in the long run, added stress is the last thing an animal with cancer needs. These experiments *cannot* be used to indicate the safety or benefit of PEMF in regard to cancer. Since to heal bones the fields are directed only at small regions, PEMF as used on humans does *not* produce stress, increased immune system response, or any concomitant antitumor activity.

Certain leads, such as the electrically injected silver, remain promising. In the 1950s and 1960s, Dr. Kenneth MacLean published some interesting work on the use of magnetic fields versus tumors in mice. He believed he'd healed several cases of cancer with steady-state magnetic fields, and certain unorthodox healers in America and India who use permanent magnets have made similar claims. The difference in effect between steady-state and time-varying fields (see Chapters 14 and 15) leads me to theorize that a steady-state magnetic field, if strong enough, may indeed halt mitosis in malignant cells.

Due to the prevailing outlook in cancer research, the key work remains to be done. Even promising nonelectromagnetic approaches have been victims of bias. There's sound evidence, for example, that megadoses of vitamin C do slow tumor growth and increase the chances for a complete cure, but Linus Pauling hasn't been able to persuade any of the powerful institutes to perform a large-scale trial. Some animal tests are now being funded, but, since vitamin C is nontoxic, immediate clinical experiments on large numbers of humans would make much more sense.

In following up the regeneration connection, two experiments in particular are crying out to be tried. Someone must attempt to duplicate the electrical environment of regeneration around tumors in lab animals, using electrodes. This would involve introducing small negative currents to thoroughly test the hypothesis that cancer cells are stuck in a state of incomplete dedifferentiation. The idea would be to dedifferentiate them the rest of the way and then let normal processes in the body turn them into healthy mature cells. The same hypothesis should be tested another way by surgically creating neuroepidermal junctions near the tumors.

Will these experiments be done soon? I wish I knew. The multi-billion-dollar cancer research bureaucracy could certainly afford them, but, although there are a few signs of change, the establishment is stuck in the near-primitive state of the war mentality. I've maintained for many years that we won't learn much more about abnormal growth until we learn more about the normal kind. That approach can lead to cancer treatments that are truly compatible with our bodies, far safer and more effective than the simplistic, dangerous ones now in vogue.

Part 4

The Essence of Life

And if the body were not the soul, what is the soul?

—WALT WHITMAN

A hallucination is merely a reality that we normally don't
have to bother with.

—STELLA DENOVA

Thirteen

The Missing Chapter

Medical students often experience a profound, ego-wrenching shock at the midpoint of their training, as the focus shifts from classroom to bedside. My experience was typical, since this division was even sharper in the 1940s than it is today. After two years' study of the scientific underpinnings of medicine, my classmates and I thought we were pretty smart. During long days and longer nights of lectures, notes, labs, books, reviews, papers, and exams, we'd drunk so deeply of the distilled wisdom of the ages that surely we must know all there was to know about bodies and diseases. All that was left, it seemed, was learning how to apply that knowledge as apprentice doctors. Then we began to study with the senior members of the clinical departments, veterans of a less scientific era, who brought us up short. Their message soon became clear: We didn't know so damn much after all; no one did. All our learning was fine as far as it went, but on the hectic wards of Bellevue things often didn't go by the book.

The greatest teacher I had in those days was my surgery professor, Dr. John Mulholland, a granite cliff of a man whose iron-gray crew cut emphasized his uncompromising idealism. Any hint of laziness, impatience, or unconcern from his students brought a brusque reprimand, but Mulholland was gravely polite and compassionate to the dirtiest wino who needed a doctor. He showed us countless problems and techniques in his demonstration-lectures held in the nineteenth-century amphitheater, but he repeated one salient message over and over: "The surgeon can cut, remove, or rearrange the tissues, and sew up the wound, but only the patient can do the healing. Surgeons must always be humble before this miracle. We must treat the tissues with sure, deft gentleness, and above all we must do no harm, for we are nothing more than nature's assistants."

None of our textbooks could tell us the how and why of healing. They explained the basics of scientific medicine—anatomy, biochemistry, bacteriology, pathology, and physiology—each dealing with one aspect of the human body and its discontents. Within each subject the body was further subdivided into systems. The chemistry of muscle and bone, for example, was taught separately from that of the digestive and nervous systems. The same approach is used today, for fragmentation is the only way to deal with a complexity that would otherwise be overwhelming. The strategy works perfectly for understanding spaceships, computers, or other complicated machines, and it's very useful in biology. However, it leads to the reductionist assumption that once you understand the parts, you understand the whole. That approach ultimately fails in the study of living things—hence the widespread demand for an alternative, holistic medicine—for life is like no machine humans have ever built: It's always more than the sum of its parts.

With delicate lab culture methods we can remove from an animal certain organs and tissues, such as bone, a heart, a pancreas, a brain, or groups of nerve cells, keeping them alive for days or weeks. Much of modern biology in the West is based on the behavior of such isolated systems, which is assumed to be the same as in the living body. Russian biology, based on Ivan Pavlov's concept of the body as an indivisible unit, has always been skeptical of tissue culture results, considering these "parabiotic" reactions only hints toward definitive studies of the entire animal. The good sense of this view is shown by the fact that life tolerates fragmentation very poorly: Except in the simplest species removal of anything more than a few cells always destroys the organization, and hence the organism. Even if we could culture separately all the organs and tissues and then put them together like Dr. Frankenstein's monster, we would still, at our present level of knowledge, have only a collection of different kinds of meat, not a living entity. As Albert Szent-Györgyi once wrote, "Biology is the science of the improbable," and seldom can we predict new discoveries from what we already understand.

These limitations were clearly recognized by American medicine in the 1940s, but they seem to have been gradually forgotten. Today most M.D.'s tacitly assume that once a few blank spots are filled in, the established basic sciences will be all we'll ever need to take care of the sick. As a result, they're losing the forest among the trees. Of the disciplines that form medicine's foundation, only physiology tries to unite structure and function into a complete picture of how the body works. Hence it's often called the queen of the biological sciences, yet even in this realm the synthesis is made organ by organ. There is, however, one organ group—the nervous system—that coordinates the activities of all the others; by receiving, transmitting, and storing information, it unites all the parts into that transcendence of fragments, the organism. Therefore the one discipline that comes closest to dealing with a living thing in its entirety is neurophysiology, which in the 1940s was already so sophisticated as to be almost a science unto itself.

Even neurophysiology couldn't explain the mystery of healing, however. My best texts either ignored it completely or shrugged it off in a few vague paragraphs. Moreover, my experiences at Bellevue during my internship and early residency convinced me that a physician's success was largely due, not to technical prowess, but to the concern he or she displayed toward patients. The patient's faith in the doctor profoundly affected the outcome of many treatments. Certain remedies, such as penicillin used against bacteria susceptible to it, worked every time. Other prescriptions weren't so predictable, however. If the patient *thought* the remedy would work, it usually did; otherwise it often didn't, no matter how up-to-date it was. Unfortunately, the importance of the doctor-patient relationship was being downgraded by the new scientific medicine. The new breed of physicians argued that this power of belief somehow wasn't real, that the patients only thought they were getting better—a bit of arrant nonsense that should have been quickly dispelled by a little open-minded, caring attentiveness on daily rounds. There was no known anatomical structure or biochemical process that provided the slightest reason to believe in such a thing, so it came to be dismissed as a mirage left over from the days of witchcraft. The placebo effect, as it's now called, wasn't documented until several decades later and still isn't fully accepted as an integral part of the healing process, but over the years I became convinced that it was a physiological effect of mind on body, just as real as the effects of wind on a tree.

Our lack of knowledge about healing in general and its psychological component in particular sowed seeds of doubt in my mind. I no longer believed that our science alone was an adequate basis for medical practice. As a surgeon, I tried to apply the principle of interaction on my own wards, by spending more time talking with patients, letting them know that I cared *for* them as well as taking care *of* them. Naturally, as I became a teacher, I tried to pass on my beliefs to others. As I gained experience, I grew more and more convinced that all the textbooks were missing a chapter—the one that should have tied it all together and helped us doctors understand the bodily harmony we were trying to restore.

When I entered research, I aimed for a fairly limited goal among the many that lured me—finding out what stimulated and controlled the growth needed for healing—but always in the back of my mind were the larger questions that had haunted me since medical school: What unified an organism, making every cell subservient to the needs of the whole? How was it that the whole being could do things that none of its components could do separately? What made an organism self-contained, self-directed, self-repairing? When you get right down to it, I wanted to know what made living things alive. Intuitively I felt sure the answers needn't be forever hidden in mystic conundrums but were scientifically knowable. However, they would require a fresh approach from science, not the simple mechanistic dogmas left over from last century. As a result of the research on nerves and regeneration described in the foregoing pages, I believe I can now sketch at least an outline of that missing chapter.

It had been known for centuries that the nerves are the body's communications lines. Still, all the information collected by neurophysiologists hadn't revealed the integrating factor behind healing. Marc Singer proved that nerves are essential for regeneration, yet the elaborate impulse and neurotransmitter system, which until recently constituted everything we knew about nerves, carries no messages during the process. Nerves are just as essential to simpler kinds of healing. Leprosy and diabetes sometimes destroy nerve function to the extremities. When this happens, a wounded limb not only fails to heal but often degenerates far beyond the actual injury. I often thought about this paradox in connection with the other realities that were poorly explained by nerve impulses, such as consciousness and its many levels, sleep, biological cycles, and extrasensory experiences. As a doctor, however, I was most concerned with the mystery of pain.

This is the least understood of sensory functions, but it must have been one of the very first to evolve. Without it, living things would be so poorly designed that they couldn't survive, for they would never know what constituted danger or when to take defensive action. Pain is quite distinct from the sense of touch. If you place your finger on a hot stove, you feel the touch first and the pain appears a discernible time later, after the reflex has already drawn your hand away. Clearly the pain is conveyed by a different means. Furthermore, there are different types of pain. Pain in the skin is different from pain in the head or belly or muscles. If you ever want to embarrass a neurophysiologist, ask for an explanation of pain.

Early in my work on regeneration it occurred to me that I'd stumbled upon another method of nerve function. I imagined slowly varying currents flowing along the neurons, their fluctuations transmitting information in analog fashion. Though I kept my main focus on the role of these currents in healing, I pursued other lines of inquiry on the side. I did so partly out of simple curiosity, but also because I realized that, no matter how much merit my DC theory might have for healing, it would have a better chance of being considered if I could fill in some of its details in a wider context.

In the study of healing I dealt only with the output side of the system, the voltages and currents sent *to* the injured area to guide cells in repairing the damage. Cybernetics and common sense alike told me that, before an organism could repair itself, it must know it had been injured. In other words, the wound must hurt, and the pain must be part of an input side of the system. Certainly if the output side was run by electrical currents, it made no sense to assume that the input side relied on nerve impulses.

At the same time another problem nagged me. The impulses and the current seemed to coexist, yet everything we knew about nerve impulses and electricity said they couldn't travel through the same neuron at the same time without interfering with each other. We now have solutions to both problems, thanks to serendipity. The ways in which the answers came show how one experiment often furthers an unrelated one, and how politics occasionally benefits science.

The Constellation of the Body

In the early 1960s, after I'd published a few research papers, I had an unannounced visitor, a colonel from the Army Surgeon General's office. He said he'd been following my work from the start and had an idea he wanted to discuss. He asked if I'd ever heard of acupuncture.

I told him it wasn't the sort of thing taught in medical school. Although I'd read about it, I had no direct experience of it and didn't know whether it did any good.

"I can tell you for sure it does work," he replied. "It definitely relieves pain. But we don't know *how* it works. If we knew that, the Army might adopt it for use by medics in wartime. After reading your work, some of us wondered if it might work electrically, the same as regeneration seems to. What do you think?"

That was a new idea to me, but right away I thought it was a good one. Although neurophysiologists had studied pain intensively for decades, there was still no coherent theory of it, or its blockage by anesthetics and anodynes. Because of Western medicine's biochemical bias, no pain-killers other than drugs were considered seriously. Maybe a physical method could give us a clue as to what pain really was.

We talked for several hours, but afterward I heard no more from the colonel, and I didn't get the chance to follow up his idea until more than a decade later. In 1971, while touring China as one of the first Western journalists admitted by the Communists, *New York Times* columnist James Reston saw several operations in which acupuncture was the only anesthetic, and he himself had postoperative pain relieved by needles after an emergency appendectomy. His reports put acupuncture in the news in a big way. It was almost the medical equivalent of Sputnik. Soon the National Institutes of Health solicited proposals for research on the Chinese technique, and I jumped at the chance.

At that time the prevailing view in the West was that if acupuncture worked at all, it acted through the placebo effect, as a function of belief. Hence it should be effective only about a third of the time, just like dummy pills in clinical tests. Many of those applying for the first grants began with this idea, and with the corollary that it wouldn't matter where you put the needles. Thus, much of our earliest research merely disproved this fallacy, which the Chinese—and apparently the U.S. Army—had done long ago. Recalling my talk with the colonel, I proposed a more elegant hypothesis.

The acupuncture meridians, I suggested, were electrical conductors that carried an injury message to the brain, which responded by sending back the appropriate level of direct current to stimulate healing in the troubled area. I also postulated that the brain's integration of the input included a message to the conscious mind that we interpreted as pain. Obviously, if you could block the incoming message, you would prevent the pain, and I suggested that acupuncture did exactly that.

Any current grows weaker with distance, due to resistance along the transmission cable. The smaller the amperage and voltage, the faster the current dies out. Electrical engineers solve this problem by building booster amplifiers every so often along a power line to get the signal back up to strength. For currents measured in nanoamperes and microvolts, the amplifiers would have to be no more than a few inches apart—just like the acupuncture points! I envisioned hundreds of little DC generators like dark stars sending their electricity along the meridians, an interior galaxy that the Chinese had somehow found and explored by trial and error over two thousand years ago. If the points really were amplifiers, then a metal needle stuck in one of them, connecting it with nearby tissue fluids, would short it out and stop the pain message. And if the integrity of health really was maintained by a balanced circulation of invisible energy through this constellation, as the Chinese believed, then various patterns of needle placement might indeed bring the currents into harmony, although that part of the treatment has yet to be evaluated by Western medical science.

The biggest problem Western medicine had in accepting acupuncture was that there were no known anatomical structures corresponding to the meridians, those live wires supposedly just under the skin. Some investigators claimed to have located tiny clusters of sensory neurons where the points were, but others had looked for them in vain. My proposal offered a convenient way into the problem. If the lines and points really were conductors and amplifiers, the skin above them would show specific electrical differences compared to the surrounding skin: Resistance would be less and electrical conductivity correspondingly greater, and a DC power source should be detectable right at the point. Some doctors, especially in China, had already measured lower skin resistance over the points and had begun using slow pulses of current, about two per second, instead of needles. If we could confirm these variations in skin resistance and measure current coming from the points, we'd know acupuncture was real in the Western sense, and we could go on confidently in search of the physical structures.

I got the grant and used part of the money to hire Maria Reichmanis, a brilliant young biophysicist who was Charlie Bachman's last Ph.D. student. Her combination of mathematical gifts and practicality got us results fast. Together we designed a "pizza cutter" electrode, a wheel that we could roll along the meridians to give us a reliable continuous reading, as well as a square grid of thirty-six electrodes to give us a map of readings around each point.

Along the first meridians Maria measured, the large-intestine and pericardial lines on the upper and lower surfaces, respectively, of each arm, she found the predicted electrical characteristics at half of the points. Most important, the same points showed up on all the people tested. Since acupuncture is such a delicate blend of tradition, experiment, and theory, the other points may be spurious; or they may simply be weaker, or a different kind, than the ones our instruments revealed. Our readings also indicated that the meridians *were* conducting current, and its polarity, matching the input side of the two-way system we'd charted in amphibians, showed a flow into the central nervous system. Each point was positive compared to its environs, and each one had a field surrounding it, with its own characteristic shape. We even found a fifteen-minute rhythm in the current strength at the points, superimposed on the circadian ("about a day") rhythm we'd found a decade earlier in the overall DC system. It was obvious by then that at least the major parts of the acupuncture charts had, as the jargon goes, "an objective basis in reality."

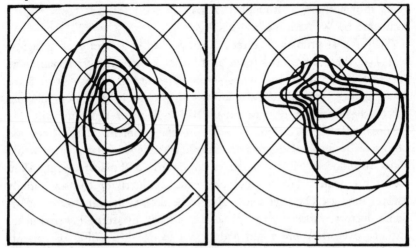

ELECTRICAL CONDUCTIVITY MAPS OF SKIN AT ACUPUNCTURE POINTS

Maria, Joe Spadaro, and I began a more sophisticated series of tests. We planned to record from six major points along one meridian as a needle was inserted into the outermost point. If the DC theory was valid, a change in potential should travel from point to point along the line. However, just as we were entering this second phase, the NIH canceled our grant, even though we'd published four papers in a year. Supposedly it had lost interest in acupuncture, at least in the kind of basic research we were doing on it. Even so, I was fairly satisfied. The input system worked as I'd predicted. The other major question remained: What structure carried the current so as not to interfere with the nerve impulses?

Of course, I've given away the answer in previous chapters. The perineural cells appear to carry the current. In the early 1970s, however, we only suspected this. The evidence came unexpectedly by cross-fertilization from an unrelated project.

One of the main problems of medical research is finding a suitable "animal model" for human diseases. The study of unmended fractures is especially hard, because people are endowed with lesser fracture-healing abilities than most other animals, for whom nonunions are a non-problem. Based on what I'd learned about the importance of nerves in bone healing, I figured we could produce nonunions in rats by cutting the nerves to the broken leg, particularly if we removed whole segments of the nerves so they couldn't grow back. I assigned this part of the project to Dr. Bruce Baker, a young orthopedic surgeon who was then finishing his residency with an extra year on a fellowship in my lab.

After Bruce had worked out the complicated surgical procedure, we anesthetized a series of rats, removed the nerve supply to one leg of each animal, and broke the fibula, or smaller bone of the calf, in a standard way. Then every day we reanesthetized a few of the rats and took out the fracture area to mount it for the microscope. At the same time, Bruce checked the cut nerves to make sure there was no regrowth. Successful denervation was confirmed by the microscope and by complete paralysis of the affected leg.

The results were encouraging yet puzzling. The nerves didn't regrow, and the broken bones took twice the normal six or seven days to heal, but heal they did, even though theoretically they shouldn't have knit at all without nerves.

It was well known that the severed end of a nerve would die after a couple of days, but, since we'd cut the nerves at the same time as we'd broken the bones, maybe the cut ends had exerted a subdued healing effect while they remained alive. In another series of animals, we cut the nerves first. Three days later, after making sure the legs were fully de-nervated, we operated again to make the fractures. We felt sure the delay would give us true nonunions. To our surprise, however, the bones healed *faster* than they had in our first series, although they still took a few days longer than normal.

Here was a first-class enigma. The only thing we could think of doing was to cut the nerves even earlier, six days before the fractures. When we got that series of slides back, we found that these animals, whose legs were still completely without nerves, healed the breaks just as fast and just as well as the normal control animals. Then we took a more detailed microscopic look at the specimens Bruce had taken from around the nerve cut. We found that the Schwann cell sheaths were growing across the gap during the six-day delay. As soon as the perineural sleeve was mended, the bones began to heal normally, indicating that at least the healing, or output, signal was being carried by the sheath rather than the nerve itself. The cells that biologists had considered merely insulation turned out to be the real wires.

Unifying Pathways

The experiments I'd done with psychiatrist Howard Friedman in the early 1960s, mentioned in Chapter 5, were the first to provide strong support for an analog theory of pain. In all animals, including humans, the normal negative potentials at the extremities weakened or vanished as an anesthetic took effect. Under deep total anesthesia, the potentials often reversed entirely, the extremities becoming positive and the brain and spine negative. At that time we didn't yet know about the two-way system, inward along sensory nerves and outward along motor nerves, but it was obvious that a current flow was being reversed by the pain-preventing drugs. In lab animals and humans under local anesthetic, such as a shot of procaine in one arm, the negative potential was abolished only for that arm. The DC potentials over the head were un-affected—except for a little blip in the recording that registered the prick of the needle!

In addition, the DC potentials react slowly enough to account for pain. A wound usually doesn't start to hurt in earnest until minutes, or even hours, after the injury. This delay has been especially hard to explain in terms of nerve impulses, which travel at 30 feet per second. However, when Friedman and I injured the limbs of salamanders while monitoring the potentials on their limbs and heads, we found that the change in the limb reading showed up in the head after a time approximating that of delayed pain. Acupuncture likewise involves a delay, usually twenty minutes or more, before its effects are felt.

We also found we could work backward, using the currents to produce anesthesia. A strong enough magnetic field oriented at right angles to a current magnetically "clamped" it, stopping the flow. By placing frogs and salamanders between the poles of an electromagnet so that the back-to-front current in their heads was perpendicular to the magnetic lines of force, we could anesthetize the animals just as well as we could with chemicals, and EEG recordings of magnetic and chemical anesthesia were identical. We got the same effect by passing a current through the brain from front to back, canceling out the normal current of waking consciousness, as in electrosleep.

One of the most exciting results of my collaboration with Dr. Friedman was proof that one's state of *waking* consciousness could change the perception of pain. Friedman, who already used hypnosis to control chronic pain in his patients, gave several of his best subjects hypnotic suggestions of arm numbness deep enough that they couldn't feel the prick of a needle. In each case, I found that the frontal negative potential of the head became less negative, often reaching zero, as the client attained deep trance. The reading changed in the same direction as in anesthesia, only not as far. Then, when the suggestion for pain control was given, the arm potential reversed just as it had in response to procaine. Conversely, when a control subject was asked in normal waking consciousness to concentrate forcefully on one arm, its sensitivity to pain *increased,* and the hand potential became more negative. We found we could use this difference to determine whether a person was really hypnotized or just cooperating.

Some doubters (including myself, I'm afraid) had believed hypnoanalgesia was merely a state in which the patient still felt the pain but didn't respond to it, but these experiments proved it was a real blockage of pain *perception.* It seems that the brain can shut off pain by altering the direct-current potentials in the rest of the body "at will." There's every reason to suppose that pain control through biofeedback or yoga likewise works by using an innate circuit for attenuating the pain signal, which releases a shot of the body's own pain-killers. When the signal is appropriately modulated, it releases endorphins (internally produced opiates), as shown by experiments in which an injection of the opiate-antagonist naloxone negates the anesthesia of acupuncture. I predict that research on this system will eventually let us learn to control pain, healing, and growth with our minds alone, substantially reducing the need for physicians.

Direct evidence for the perineural DC system has been accumulating gradually for several decades. Electric currents were detected in the glial cells of rat brains as long ago as 1958, and good (though long-ignored) measurements of direct currents in the frog's brain go back to the work of Ralph Gerard and Benjamin Libet in the early 1940s. Electron microscope work has shown that the cytoplasm of all Schwann cells is linked together through holes in the adjacent membranes, forming a syncytium that could provide the uninterrupted pathway needed by the current. The other perineural cells—the ependyma and glia—are probably connected in the same way, for syncytial links have recently been found in the glia of the leech, whose nervous system is much studied because of its unusually large cells. Recent use of selective radiation to isolate Schwann cells has shown that they, and not the neuron fibers, supply the nerve stimulus essential to regeneration.

The invention of a better magnetometer has yielded definitive proof that's now widely acknowledged. Any electric current automatically generates a magnetic field around itself. Hence, as the perineural current conveys information in its fluctuations, it must be reflected by a magnetic field around the body, whose pulsing would reveal the same information. When I first proposed this idea, many of my colleagues dismissed it as rank nonsense. I couldn't prove them wrong, because there were no instruments to measure a field as weak as that generated by such small currents. Everyone knew the human body had no effect on a compass needle or any other magnetic-field detector available at the time.

Then, in 1964, a solid-state physicist named Brian D. Josephson invented the electronic device now called a Josephson junction, a simple item that won him a Nobel Prize. Basically it consists of two semiconductors connected so that current can oscillate in a controlled fashion between them. Today it has many applications, especially in computers. When cooled near absolute zero in a bath of liquid helium, it becomes a *super*conductor in which the current plays back and forth endlessly. Superconduction is the passage of electrons through a substance without the resistance normally found in any conductor. This apparatus, called a superconducting quantum interferometric device, or SQUID for short, is a magnetic field detector thousands of times more sensitive than any previously known.

In 1963, G. M. Boule and R. McFee just barely managed to measure the relatively large magnetic field produced by the human heart—using the best old-fashioned instrument, a coil with 2 million turns of wire. Then, in 1971, working in a null-field chamber, from which the earth's magnetism and all artificial fields were screened out, Dr. David Cohen of MIT's Francis Bitter National Magnet Laboratory, who'd been corresponding with our lab since the early years, first used the SQUID to measure the human head's magnetic field. Two kinds of magnetic fields have been found. Quickly reversing AC fields are produced by the back-and-forth ion currents in nerve and muscle. They're strongest in the heart, since its cells contract in synchrony. The SQUID has also confirmed the existence of the direct-current perineural system, which, especially in the brain, produces steady DC magnetic fields one billionth the strength of earth's field of about one-half gauss.

By 1975, Drs. Samuel Williamson, Lloyd Kaufman, and Douglas Brenner of NYU had succeeded in measuring the head's field without a shielded enclosure, even amid the electromagnetic noise of downtown Manhattan. More important, they've found that the magnetoencephalogram (MEG)—a recording of changes in the brain's field analogous to the EEG—is often a more accurate reflection of mental activity than the EEG. Because the magnetic field passes right through the dura, skull bones, and scalp without being diffused, an MEG locates the current source more accurately than EEG measurements. The NYU group has since begun correlating magnetic events with well-known cerebral responses, such as the reaction of cells in the visual cortex to simple patterns and flashes of light. When the brain reacts to any stimulus, it produces a wave of electrical activity that's contained in the EEG. It's invisible in a standard EEG recording because so much else is always going on in the brain at the same time. However, when one simple stimulus is repeated many times and the EEG tracings are averaged by computer, the particular electrical response to that one stimulus—called an evoked potential—can be teased out. Several research groups have slowly built up a small vocabulary of wave forms with specific meanings, including a "surprise wave," an "intention wave," and a "double-take wave," which appears when the mind briefly tries to make sense of semantic nonsense, as in the statement "She took a drink from the radio."

Taken together, the MEG research so far seems to be establishing that every electrical evoked potential is accompanied by a magnetic evoked potential. This would mean that the evoked potentials and the EEG of which they're a part reflect true electrical activity, not some artifact of nerve impulses being discharged in unison, as was earlier theorized. Some of the MEG's components *could* come from such additive nerve impulses, but other aspects of it clearly indicate direct currents in the brain, particularly the central front-to-back flow. The MEG doesn't show the EEG's higher-frequency components, however, suggesting that some parts of the two arise from different sources.

Since every reaction and thought seems to produce an evoked potential, the DC system seems directly involved in every phase of mental activity. At the very least, the electric sheath acts as a bias control, a sort of background stabilizer that keeps the nerve impulses flowing in the proper direction and regulates their speed and frequency. But the analog structure probably plays a more active role in the life of the mind. Variations in the current from one place to another in the perineural system apparently form part of every decision, every interpretation, every command, every vacillation, every feeling, and every word of interior monologue, conscious or unconscious, that we conduct in our heads.

This part of the analog system's job is much less well understood, however, than its integrative function throughout the rest of the body. Perineural cells accompany every part of the nervous system. Even the tiniest twiglets of sensory nerves in the skin, which don't have a myelin covering, are surrounded by Schwann cells. The perineural structures are thus just as well distributed to integrate bodily processes as the nerves themselves. They reach into each area of the body to create a normal electrical environment around each cell, or a stimulatory one when healing growth is needed. Likewise they enable an organism to sense the type and extent of damage anywhere in the body by transmitting the current of injury, with its by-product of pain, to the CNS. One could take a "fantastic voyage" from the farthest Schwann cell outpost in the big toe through the spinal cord and into all parts of the brain. Indeed, electrons are making this trip every moment of our lives.

Thus our bodies have an intricate and multilayered self-regulating feedback arrangement. We know, on the psychological level, that a person's emotions affect the efficiency of healing and the level of pain, and there's every reason to believe that emotions, on the physiological level, have their effect by modulating the current that directly controls pain and healing.

These discoveries give us a testable physical basis for the placebo effect and the importance of the doctor-patient bond. They also may give us the key to understanding the "miracle" cures of shamans, faith healers, and saints, as well as the spontaneous healing reported by means of visions, prayer, yoga, or battlefield terror. At the Menninger Foundation, Elmer Green has long been using biofeedback to explore the mind-body relationship. Green has described the full yogic control of pain and healing developed by one of his subjects, an otherwise average businessman. He lay against a bed of nails with no pain, and, when informed that a puncture from one of the points was bleeding, he turned his head, gazed at the wound, and immediately stopped the flow. A combination of biofeedback, recording electrodes, and the SQUID magnetometer would seem to be the ideal setup for the next level of inquiry into the mind's healing powers.

Moreover, since the analog system, like the impulse network, appears to work on both conscious and subconscious levels simultaneously, it's a likely missing link in several other poorly understood integrative functions that also cross from one realm to the other. It may lead us at last to fathom the twin wells of memory and emotion. It may even help us understand what happens when a new synthesis of creative thought, a.k.a. inspiration, bursts forth like a mushroom from strands of mycelia that have been quietly gathering their subterranean forces. Then science for the first time will begin to comprehend the artistic essence that makes its rational side productive.

Fourteen

Breathing with the Earth

Major changes in one's life often proceed improbably from the most minor events. So it was that I became involved in one of the most interesting parts of my work in 1961 because a dog bit someone I didn't even know at the time.

My first electrical measurements on salamanders had just revealed part of the DC control system to me. Elementary physics told me that the currents and their associated electromagnetic fields would have to be affected in some way by external fields. In engineering terms, the biomagnetic field would be coupled to the DC currents. Hence changes impressed upon it by external fields would be "read out" through perturbations in the current. Outside fields would also couple directly to the currents themselves, without acting through the biofield as intermediary, especially if the currents were semiconducting. In short, all living things having such a system would share the common experience of being plugged in to the electromagnetic fields of earth, which in turn vary in response to the moon and sun. In the late eighteenth century, the Viennese hypnotist and healer Franz Anton Mesmer proposed direct magnetic influences on earthly bodies from the heavenly ones, but his idea came from the scientifically unacceptable domain of astrology. With the notable exception of Nikola Tesla, most prominent researchers have derided it until recently. I figured the DC system must be the missing link in a very different, but very real, connection between geophysics and the responses of living things. I was eager to investigate it, but at first I didn't know how.

I was an orthopedic surgeon, about as far removed as possible from the psychiatric expertise needed for a serious study of behavior. And suppose I did find something? Who would believe me if I ventured so far from my specialty? The whole idea was preposterous to the science of the time, anyway. Still, I had to do something.

During the International Geophysical Year of 1957–58, I'd been a volunteer in the Aurora Watch Program. To find out whether the northern lights appeared simultaneously throughout the north latitudes in response to changes in the earth's magnetic field (they did), IGY organizers recruited a worldwide network of amateur observers to go out into their backyards every night and look at the sky. All of us got weekly reports on the state of the field from the national magnetic observatory at Fredericksburg, Virginia. I decided to go back through this data and see if there was any correlation between the disturbances in earth's field caused by magnetic storms on the sun, and the rate of psychiatric admissions to our VA hospital.

Luckily for me, Howard Friedman, the hospital's chief of psychology, was collecting donations door to door for a local Boy Scout troop at about this time. At one house, the family dog took an instant dislike to him and bit his ankle. After bandaging the wound, Howard's doctor gave him a tetanus booster shot. As luck would have it, Howard came down with a rare allergic reaction that involved fever, fatigue, nausea, and painful swelling of all the joints.

Since I was the nearest bone-and-joint man, Howard came to see me. This type of reaction is frightening, but not dangerous, and disappears of its own accord in a day or two. After I made the diagnosis and reassured him, we sat and talked for a few minutes. After some chitchat about the shortcomings of the hospital administration, he gestured at the papers tacked all over the walls of my office and asked, "What are all those charts?" I told him about my magnetic brainstorms.

He obviously thought I was as crazy as the people whose admissions I was charting, and probably wondered about the advice I'd just given. However, after hearing the background, he agreed it wasn't as silly as it sounded, and offered to help. It was a real break for me, since he was already a respected researcher, and a practical, open-minded one to boot. My diagnosis was correct, and our collaboration lasted almost two decades.

Howard's reputation got us access to the records of state psychiatric hospitals, giving us a sample large enough to be statistically useful. We matched the admissions of over twenty-eight thousand patients at eight hospitals against sixty-seven magnetic storms over the previous four years. The relationship was there: Significantly more persons were signed in to the psychiatric services just after magnetic disturbances than when the field was stable. Of course, such a finding could only serve as a guide to further investigation, because so many factors determined whether a person sought psychiatric help, but we felt that other influences would even out over such a large number of patients.

Next we looked for the same type of influence in patients already hospitalized. We selected a dozen schizophrenics who were scheduled to remain in the VA hospital for the next few months with no changes in treatment. We asked the ward nurses to fill out a standard evaluation of their behavior once every eight-hour shift. Then we correlated the results with cosmic ray measurements taken every two hours from government measuring stations in Ontario and Colorado. Since magnetic storms were generally accompanied by a decrease in the cosmic radiation reaching earth, we thought we might find changes in the patients' actions and moods during these declines. We decided to use cosmic rays instead of direct reports of the magnetic field strength because of problems in distinguishing between magnetic storms and other variations in the earth's field.

The nurses reported various behavior changes in almost all the subjects one or two days after cosmic ray decreases. This was a revealing delay, for one type of incoming radiation—low-energy cosmic ray flares from the sun—was known to produce strong disruptions in the earth's field one or two days later.

With this encouragement we went on in 1967 to confirm, by experiments described more fully in the next chapter, that abnormal magnetic fields did produce abnormalities in various human and animal responses. We found slowed reaction times in humans and a generalized stress response in rabbits exposed to fields ten or twenty times the normal strength of the earth's. Hence we suspected that the earth's normal field played a major role in keeping the DC system's control of bodily functions within normal bounds. The proof of this idea has come mainly from the work of two men: Frank Brown at Northwestern and Rutger Wever, working at the Max Planck Institute in Munich.

Already a respected endocrinologist, Brown became interested in biocycles in the 1950s. It was common knowledge that most organisms had a circadian rhythm of metabolic activity, which most people assumed was directly linked to the alternation of night and day or, in the case of shore life, to the tides. Oysters, for example, would open their shells to feed whenever the tide came in, covering them with water. It was a simple, obvious observation, but Brown didn't take it for granted. To his surprise, oysters in an aquarium with constant light, temperature, and water level still opened and closed their shells in time with their compatriots at the beach. To find out why, Brown flew oysters in a lightproof box from New Haven to his lab, in Evanston, Illinois. At first they kept time with Connecticut oysters, then in a few weeks gradually shifted to the tide pattern Evanston would have had if it had been on a seacoast. The oysters not only knew they'd been taken 1,000 miles westward, they also suffered from jet lag!

In his search for a creature whose response to magnetic fields might tell him more about biocycles, Brown settled upon the mud snail *Nassarius,* at home in the intertidal zone anywhere in the world. In his lab he placed the snails under uniform illumination in a box with an exit facing magnetic south. When they left the enclosure in early morning, they turned west. When leaving at noon, they turned east, but took a westerly course again in early evening. Furthermore, at new and full moon, the snails' paths veered to the west, while at the quarters they tended more eastward than at other times.

Brown's precise data from this and many other experiments showed that *Nassarius* had two clocks, one on solar time and one on lunar, and subsequent work with magnets told something about how the timepieces ran. The earth's magnetic field averaged 0.17 gauss in Evanston. When Brown put a 1.5 gauss permanent magnet facing north-south underneath the snails' doorways to augment the natural field, the animals made sharper turns, but their direction wasn't affected. Turning either the magnet or the enclosure through various angles made the snails change course a specific number of degrees. Brown concluded: "It seemed as if the snails possessed two directional antennae for detecting the magnetic field direction, and that these were turning, one with a solar day rhythm and the other with a lunar day one." This crucial experiment not only showed the dependence of biocycles on the earth's magnetic field, it also demonstrated the subtlety of the link. No longer could we expect changes in the magnetic environment to be as obvious in their effects on life as changes in oxygen levels, food supply, or temperature.

The niceties of the earth's electromagnetic field itself became better known as Brown's work progressed. Far from a static, simple magnetic field like that around a uniform bar of magnetized iron, the earth's field has turned out to have many components, each full of quirks.

At the end of the nineteenth century, geophysicists found that the earth's magnetic field varied as the moon revolved around it. In the same period, anthropologists were learning that most preliterate cultures reckoned their calendar time primarily by the moon. These discoveries led Svante Arrhenius, the Swedish natural philosopher and father of ion chemistry, to suggest that this tidal magnetic rhythm was an innate timekeeper regulating the few obvious biocycles then known.

Since then we've learned of many other cyclic changes in the energy structure around us:

- The earth's electromagnetic field is largely a result of interaction between the magnetic field per se, emanating from the planet's molten iron-nickel core, and the charged gas of the ionosphere. It varies with the lunar day and month, and there's also a yearly change as we revolve around the sun.
- A cycle of several centuries is driven from somewhere in the galactic center.
- The earth's surface and the ionosphere form an electrodynamic resonating cavity that produces micropulsations in the magnetic field at extremely low frequencies, from about 25 per second down to 1 every ten seconds. Most of the micropulsation energy is concentrated at about 10 hertz (cycles per second).
- Solar flares spew charged particles into the earth's field, causing magnetic storms. The particles join those already in the outer reaches of the field (the Van Allen belts), which protect us by absorbing these and other high-energy cosmic rays.
- Every flash of lightning releases a burst of radio energy at kilocycle frequencies, which travels parallel to the magnetic field's lines of force and bounces back and forth between the north and south poles many times before fading out.
- The surface and ionosphere act as the charged plates of a condenser (a charge storage device), producing an electrostatic field of hundreds or thousands of volts per foot. This electric field continually ionizes many of the molecules of the air's gases, and it, too, pulses in the ELF (extremely low frequency) range.
- There are also large direct currents continually flowing within the ionosphere and as telluric (within-the-earth) currents, generating their own subsidiary electromagnetic fields.
- In the 1970s we learned that the sun's magnetic field is divided from pole to pole into sectors, like the sections of an orange, and the field in each sector is oriented in the direction opposite to adjacent sectors. About every eight days the sun's rotation brings a new region of the interplanetary (solar) magnetic field opposite us, and the earth's field is slightly changed in response to the flip-

flop in polarity. The sector boundary's passage also induces a day or two of turbulence in earth's field.

The potential interactions among all these electromagnetic phenomena and life are almost infinitely complex.

For many years most scientists dismissed Brown's conclusions as impossible. Given the old premise that life was entirely a matter of water chemistry, none of these electromagnetic changes would have enough energy to affect an organic process in any way. Discovery of the DC system showed how the interaction could work without energy transfer; it gave living things a way of "sensing" the fields directly. Undaunted by the slow acceptance of his work, Brown went on to document *Nassaria*'s sensitivity to electrostatic fields as well. He also found magnetically driven cycles in all other organisms he tested, including mice, fruit flies, and humans. Even potatoes in a bin showed a field-linked rhythm of oxygen consumption. In humans, hormone output and the number of lymphocytes in the bloodstream are but two of many variables that dance to the same beat. One of the most important is cell cycle time. The actual process of cell division—in which chromosomes appear, line up, split in half, and are distributed equally between the two cells—takes only a few minutes. It must be preceded by several longer stages, one of which is duplication of all the cell's DNA. All stages together take *about one day*. Thus all growth and repair, which depend on regulated cell division, are synchronized with the earth's field.

Rutger Wever has done some even more telling work with humans during the last decade and a half. He built two underground rooms to completely isolate people from all clues to the passage of time. One was kept free of outside changes in light, temperature, sound, and such ordinary cues, but wasn't shielded from electromagnetic fields. The other room was identical but also field free. Observing several hundred subjects, who lived in the bunkers as long as two months, and charting such markers as body temperature, sleep-waking cycles, and urinary excretion of sodium, potassium, and calcium, Wever found that persons in both rooms soon developed irregular rhythms, but those in the completely shielded room had significantly longer ones. Those still exposed to the earth's field kept to a rhythm close to twenty-four hours. In some of these people, a few variables wandered from the circadian rate, but they always stabilized at some new rate in harmony with the basic one— two days instead of one, for example. People kept from contact with the earth's field, on the other hand, became thoroughly desynchronized. Several variables shifted away from the rhythms of other metabolic systems, which had already lost the circadian rhythm, and established new rates having no relationship to each other.

Wever next tried introducing various electric and magnetic fields into his completely shielded room. Only one had any effect on the amorphous cycles. An infinitesimal electric field (0.025 volts per centimeter) pulsing at 10 hertz dramatically restored normal patterns to most of the biological measurements. Wever concluded that this frequency in the micropulsations of the earth's electromagnetic field was the prime timer of biocycles. The results have since been confirmed in guinea pigs and mice. In light of this work, the fact that 10 hertz is also the dominant (alpha) frequency of the EEG in all animals becomes another significant bit of evidence that every creature is hooked up to the earth electromagnetically through its DC system. Recently a group under Indian biophysicist Sarada Subrahmanyam reported that the human EEG not only responded to the micropulsations, but responded differently depending on which way the subject's head was facing in relation to the earth's field. Oddly enough, however, the head direction had no effect if the subject was a yogi.

The relationship has been conclusively proven by recent studies of the pineal gland. This tiny organ in the center of the cranium has turned out to be more than the vaguely defined "third eye" of the mystics. It produces melatonin and serotonin, two neurohormones that, among many other functions, directly control all of the biocycles. The lamprey, akin to the ancestor of all vertebrates, as well as certain lizards, has an actual third eye, close to the head's surface and directly responsive to light, instead of the "blind" pineal found in other vertebrates. The eminent British anatomist J. Z. Young has recently shown that this organ controls the daily rhythm of skin color changes that these animals undergo.

For our story the most important point is that very small magnetic fields influence the pineal gland. Several research groups have shown that applying a magnetic field of half a gauss or less, oriented so as to add to or subtract from the earth's normal field, will increase or decrease production of pineal melatonin and serotonin. Other groups have observed physical changes in the gland's cells in response to such fields. The experiments were controlled for illumination, since it has been known for several years that shining a light on the head somehow modifies the gland's hormone output even though it's buried so deeply within the head in most vertebrates that, as far as we know, it can't react directly to the light.

We likely have yet to discover many other ways that energy cycles in the solar system affect life on earth. They may strongly affect the outbreak of disease, for example. The last six peaks of the eleven-year sunspot cycle have coincided with major flu epidemics. A Soviet group under Yu. N. Achkasova at the Crimean Medical Institute, working with astronomer B. M. Vladimirsky of the Crimean Observatory, has found a connection between the sun's magnetic field and the *Escherichia coli* bacteria that live in our intestines and help us digest our food. The Russians found the bacteria grew faster when the sun's field was positive, or pointing toward earth, and slowed down when it was negative. Two days after the passage of each sector boundary there was a dip in bacterial growth corresponding to the maximum geomagnetic turbulence. The data also showed a decline in growth in response to large solar flares. Other Russian scientists have drawn a tentative correlation between the sector cycle and reports from two groups of persons with neurological diseases. The patients felt worse within sectors of positive polarity, when bacteria seemed to grow faster. Life's geomagnetic coupling to heaven and earth is apparently more like a web than a simple cord and socket.

The Attractions of Home

An animal's biocycles must be appropriate for its environment if it's to survive, so they must be precisely tuned to its geographic location. We might suspect, therefore, that many creatures would use magnetic information for their sense of place. A great deal of recent work has shown that they do. A built-in compass helps guide them in foraging or other local business, as well as migration over much longer routes. The latter feats often rival those of any modern navigator. Monarch butterflies travel from Hudson Bay to South America straight across the Caribbean without ever getting lost. The arctic tern breeds in summer on the northern ice cap, then moves to Antarctica for the other hemisphere's summer, flying 11,000 miles each way. Some salamanders, only inches long and built very low to the ground, travel up to 30 miles of rugged mountain country in California to set up housekeeping, then return to their home stream to breed. Such activities are hard to experiment with, however, so we've learned more about day-to-day travel.

Karl von Frisch was the first to attack the problem, with his famous 1940s studies of the honeybee dance, which won him a Nobel Prize in 1973. He established that on clear days bees navigated by combining the sun's angle with their sense of time, the way Boy Scouts are taught to use wristwatches as compasses. The bees also had a polarized light system that could determine the sun's direction through light clouds or forest canopy. Even more amazing, Frisch found, the scouts told workers back at the hive where the flowers were by means of a dance using the sun's angle and the direction to earth's center (the gravity vector) as references. However, Frisch noted that bees could still navigate between food and home just as well on completely overcast days, when the sun's angle and polarized light weren't available. There had to be a backup system.

It soon turned out that homing pigeons had the same abilities. In 1953 G. Kramer inferred that the birds must have a compass in addition to a map of remembered landmarks from the way they immediately pointed their beaks toward home after circling once after release. Soon others found the same kind of sun compass as bees used, but the pigeons could also steer perfectly on cloudy days. Back in 1947, H. L. Yeagley had had the temerity to suggest in the *Journal of Applied Physics* that pigeons might have a magnetic sense that allowed them to use the earth's field just as we use magnetic compasses. He was ridiculed and "refuted" by a few inadequate experiments—such as placing a pigeon in a variety of electromagnetic fields and noting that it seemed to be comfortable! Others, including Yeagley, attached small magnets to the birds' heads or wings, but found no clear-cut changes in their flight patterns.

Only a few researchers quietly looked into the matter further. After Hans Fromme of the Frankfurt Zoological Institute noted in the late 1950s that caged European robins faced longingly in their normal southwest migratory direction even when they were kept from seeing sun and stars, their usual signposts, his co-worker Friedrich Merkel discovered that, insulated from the earth's field by a steel cage, they no longer faced in one particular direction. Furthermore, by changing the orientation of the surrounding field with coils, he could give the birds a false sense of where southwest was. The experiment was validated with indigo buntings several years later.

There the matter rested until 1971, when William T. Keeton of Cornell realized that the pigeon's magnetic sense, if it existed, would be overshadowed by its sun compass, so naturally magnets affixed to the birds would have no effect on *clear* days. He soon found that the same birds released on a cloudy day got lost.

To study this magnetic interference in any weather, Keeton made translucent contact lenses for his birds, then released them in the mountains of northern New York. Simulating dense clouds, the contacts blocked the sun's angle and polarized light, and with 1-gauss magnets attached to their heads the birds couldn't find their way home. However, each avian Ulysses who wore the lenses but no magnets faultlessly navigated the 150 miles southwest to Ithaca, then flew ever tighter circles around the loft and fluttered in like a helicopter to a perfect blind landing.

Carrying the work forward after Keeton's untimely death soon after this experiment, Charles Walcott and Robert Green of the State University of New York at Stony Brook, working with James L. Gould of Princeton, outfitted pigeons with miniaturized electromagnetic coils that let the researchers vary the type and orientation of applied field at will. They discovered that if the south pole of the field was directed up, the birds could still find home, but with the north pole up they flew directly *away* from it. That meant they were using magnetic north as a reference point. At about the same time two German scientists, Martin Lindauer and Herman Martin, analyzed half a million bee dances and found a "magnetic error" in them—a compensation for the difference between magnetic north and true north. They were also able to introduce specific angles of error in the dances with specifically oriented coils around the hive. Here was proof that magnetic guidance systems existed in both the birds and the bees.* The next question was how the systems worked.

In 1975 Richard P. Blakemore, then a graduate student at the University of Massachusetts in Amherst, astonished the world of biology with the announcement that some bacteria, the lowliest of all cells, also had a magnetic sense. Blakemore made the discovery when, studying the salt marshes of Cape Cod, he noticed that one type of bacterium always oriented itself north-south on his microscope slides. Soon he found magnetotactic bacteria (those reactive to magnetism) near Cambridge, Massachusetts, where he set about studying them with Richard Frankel of MIT's magnet lab. The direction to magnetic north points through the earth somewhat down from the horizon, and the scientists became convinced that the bacteria were using the field to guide themselves ever downward to the mud where they throve, since they were too small to sink through the random molecular motion of the water around them.

---

*Recent work by Cornell biologist Kraig Adler showed that the magnetic sense of salamanders was many times more acute than even that of pigeons. Not only could the amphibians find home without light or other common cues; in addition, when Adler tried to confuse them with artificial fields, they quickly adapted to the interference and oriented themselves correctly in relation to the weaker geomagnetic background.

This idea was later confirmed by findings that microbes at Rio and in New Zealand were south-seekers.

Blakemore's electron micrographs soon revealed a surprising structure. Each bacterium contained within it, like a chain of cut jet stones, a straight line of magnetite microcrystals. Surrounded by a thin membrane, each of these particles was a single domain, the smallest piece of the mineral that could still be a magnet.

Blakemore's bacteria led Gould to look for similar crystals in bees and pigeons. Since an electron microscope survey of even a bee's brain would take several lifetimes, he examined the insects with a SQUID magnetometer. After confirming that they were magnetic, he dissected them and narrowed the location down to a part of the abdomen. Using the same method, Walcott and Green dissected the heads of two dozen pigeons, gradually subdividing them with nonmagnetic probes and scalpels. After a painstaking search the investigators found a tiny magnetic deposit in a 1- by 2-millimeter piece of tissue richly festooned with nerves, on the right side of the head, between the brain and the inner table of the skull. The same dot of tissue contained yellow crystals of the iron-storage protein ferritin, indicating that the pigeons, like the bacteria, synthesized their own lodestone crystals.

As usual, these recent answers have raised plenty of new questions. The existence of magnetic sensors in such diverse creatures as bacteria, bees, and birds—the current count of species with magnetic organs is twenty-seven, including three primates—suggests that a magnetic sense has existed from the very beginning of life, perhaps only to be perfected by creatures that need to get around a lot. Do all animals, then, have the same sensors, and do they always serve the same function? How is the information read out of the crystals by the nervous system and translated into directions? What aspect of the earth's field do these organs sense?

Keeton noticed an especially odd thing about his pigeons' flight patterns. When flying on visual flight rules by sun compass, they would circle once, get their bearings, then move off straight toward Ithaca. But when using their magnetic compass, the birds would fly due west from their release point until they got out over Lake Ontario, due north of Ithaca. Then, out of sight of land, they would make a right-angle turn to the left and follow the exact meridian of home. Keeton told me this but never published the observation because he didn't know what to make of it. He said, "I asked a physicist: 'Are they making contact with a certain magnetic line of force?' The man said, 'No, magnetic lines of force are just an arbitrary convention we use to symbolize a field and describe anomalies in it, such as occur around iron ore deposits. As far as we know, the lines have no equivalent in reality, and if they did, they would vary all over the place as the earth's field changes, anyway.'" Do pigeons then follow some maplike structure in the earth's field itself, a grid like that described by dowsers and geomancers since ancient times, something we can't find today even with our SQUID? Some migrating birds make a dogleg to the east in their north-south flyway, sailing out of sight of land over Lake Superior. Do they go out of their way to avoid being disoriented by iron ore deposits in the Mesabi Range? We can suspect, but we don't yet know.

To most people, of course, the most interesting questions concern themselves. Do we, too, have compasses in our heads?

On June 29, 1979, R. Robin Baker, a young University of Manchester researcher into bionavigation, led a group of high school students into a bus at Barnard Castle, near Leeds, England. Baker blindfolded and earmuffed them, then gave them all headbands. Half of the headgear contained magnets and half contained brass bars that their wearers *thought* were magnets. As the volunteers leaned back to concentrate, Baker wove a mazelike course through the town's tangled streets, then traveled a straight highway to the southwest. After a few miles the coach stopped while the students wrote on cards an estimate of the compass direction toward the school. Then the driver turned through 135 degrees and continued east to a spot southeast of the school, where the students again estimated their direction. When the cards were analyzed, it turned out that the persons with brass bars by their heads had been able to sense the proper heading quite reliably, while those wearing magnets had not.

Gould and his Princeton co-worker K. P. Able recently tried to replicate Baker's experiment but failed. However, Baker's review of the attempt suggested that the volunteers' directional sense may have been thrown off by magnetic storms, weak magnetic gradients inside the bus, and/or the greater electromagnetic contamination of Princeton compared with rural England. Baker and co-worker Janice Mather have recently devised a simpler test method. In the middle of a specially built, light-tight wooden hut free of magnetic interference, the subject is blindfolded, earmuffed, and seated on a friction-free swivel chair. After being turned around several times, the subject must estimate his or her compass heading as before. With statistically consistent success in more than 150 persons, Baker believes he has proven the existence of a human magnetic sense.

Oddly enough, he finds that as long as people can't feel the sun or sense some other obvious cue, they can judge direction *better* with blindfolds on. Otherwise they start rationalizing the process, trying to deduce the right way from too little evidence and becoming confused. He surmises that the magnetic sense serves its purpose best by unconsciously giving a continuous sense of direction without its owner's having to be aware of it all the time, and thus freeing attention for the search for food, a mate, shelter, and so on.

In 1983, using magnetic measurements in selective-shielding experiments, Baker and his co-workers reported locating magnetic deposits close to the pineal and pituitary glands in the sinuses of the human ethmoid bone, the spongy bone in the center of the head behind the nose and between the eyes. It's interesting to note that selective-shielding studies done in the early 1970s, by Czech émigré biophysicist Zaboj Harvalik, an adviser to the U.S. Army Advanced Material Concepts Agency, pointed to this same spot as one of two areas—the other was the adrenal glands—where the dowsing ability resided.

In 1984 a group headed by zoologist Michael Walker of the University of Hawaii in Honolulu isolated single-domain magnetite crystals from a sinus of the same bone in the yellowfin tuna and Chinook salmon. The crystals were of a shape normally shown only by magnetite synthesized by living things rather than geological processes. Abundant nerve endings entered the magnetic tissue, and the crystals were organized in chains much like those in magnetotactic bacteria. Each crystal was apparently fixed in place but free to rotate slightly in response to external magnetic forces. Calculations showed that such chains would be able to sense the earth's magnetic field with an accuracy of a few seconds of arc, or a few hundred feet of surface position. This result correlated perfectly with earlier homing studies on live tuna by the same group. This detailed work, along with related earlier research, strongly suggests that all vertebrates have a similar magnetic organ in the ethmoid sinus area, and I suspect that this organ also transmits the biocycle timing cues from the earth field's micropulsations to the pineal gland.

The Face of the Deep

DNA pioneer Erwin Chargaff has called the origin of life "a subject for the scientist who has everything," but that hasn't stopped many of us (or even him, for that matter) from speculating about it. There are numerous detailed pictures of that primal scene in print today, but most are variations on one theory—"warm soup and lightning."

Life on earth began some 4 billion years ago, or roughly 1 or 2 billion years after the world itself was born. The atmosphere then was completely different, much like Jupiter's today, mostly ammonia and methane. Into that atmosphere some source of energy—lightning, heat, and ultraviolet radiation have all been suggested—led to the spontaneous formation of simple organic compounds. Sifting into the oceans for millions of centuries, these compounds would have coalesced by chance into ever more elaborate patterns. According to the theory, this process culminated in closed "protocells" able to resist the reactivity of other structures while growing through incorporation of similar compounds.

This idea owes its dominance largely to an important experiment made by S. L. Miller in 1953. Miller pumped a facsimile of the presumed early atmosphere—ammonia, methane, and water vapor—continuously past an electric spark. After several days he had some amino acids. Since these are the bricks of DNA, RNA, and all proteins, the evidence seemed very good. Later runs yielded even more sophisticated molecules. In water they coalesced into globules with a sort of membrane around them—called "coacervates" by A. I. Oparin and "proteinoids" by Sidney Fox, two of the most assiduous students of biogenesis.

Nothing came close to being alive in any of these spark chambers, however. More important, the experiments raised two difficulties, one theoretical, one practical. The soup theory needed to come up with a very sophisticated entity, a living cell with some genetic system using DNA or RNA, right off the bat. According to our notions of biology, nothing could be alive before that point, yet it seemed incredible that chance associations of the building blocks could form a palace of such complexity without passing through a mud-hut stage.

When the warm soup theory was first advanced, the mechanistic persuasion was at its height. Living things were complicated machines, but molecular machinery they were. However, the concept of a cell was much simpler than it is today. No longer is it considered a mere baglet of jelly with a few master molecules telling it what to do. Even the membrane of the simplest bacterium responds in intricate ways to information from outside, yet our best electron microscopes haven't yet revealed a complexity of structure adequate to explain its work.

There was a basic chemical problem as well. All organic compounds exist in two forms, or isomers. Each contains the same number and type of atoms, but in one they're arranged as a mirror image of the other. One is "right-handed" and the other is "left-handed." They're identified by the way they bend light in solution. The dextrorotatory (D) forms rotate it to the right, while levorotatory (L) isomers refract it to the left. All artificial methods of synthesizing organic compounds—including the spark experiments—yield a roughly equal mixture of D and L molecules. However, all living things consist of *either* D or L forms, depending on the species, but *never both*.

We must conceive of the first living things as something unexpected, not just simpler versions of what we see around us. They couldn't have been cells; they couldn't have had a DNA-RNA-protein system, a living membrane, or a nerve impulse network.

We can try to define the bare minimum, the processes that must be available before an entity can be called living. There must be a way to receive information about external conditions, process it, and store it so that the data change the being's response to the same stimulus in the future. In other words, a sort of crude consciousness and memory must be present from the first. A life-form must also be able to sense damage and repair itself. Third, we can expect that it would show some sort of cyclic activity, perhaps tuned primarily to the circadian rhythm of the lunar day. Self-replication, one of the main requirements in the DNA-based theory, can be dispensed with. An organism that can fully heal its injuries is theoretically immortal. The criteria for life can be summarized as organization, information processing, regeneration, and rhythm.

The funny thing is that all of these criteria are met by the activities of semiconducting crystals. Semiconductivity occurs naturally in several inorganic crystals, including silicon, one of the most common elements, and the rare earth germanium. Moreover, extremely small amounts of contaminants will change the crystal's electrical properties dramatically in doping. The earth's volcanic mixing would have produced minerals with a wide variety of current-handling abilities to start from. Most important, the piezoelectric, pyroelectric, photoelectric, and other responses of semiconducting crystals could have served as an analog method of processing and storing information about pressure, heat, and light. Moreover, repeated passage of current through some semiconductors *permanently* changes the materials' characteristics so as to make the same electrical responses easier in the future. Movement of electrons along the crystal lattices inevitably would have been shaped by geo-celestial cycles in the earth's electromagnetic field, as well as by the fields around other such crystalline organisms nearby—providing a sense of time and information about the neighbors. The currents also would have instantly reflected any loss of material and guided the deposition of replacement atoms to restore the original structure.

The idea of certain rocks, in the course of a billion years or so, gradually becoming responsive to their surroundings, growing, learning to "hurt" when a lava flow or sulfuric rain ate away part of a vertex, slowly rebuilding, pulsing with, well, life, even developing to a liquid crystal stage and climbing free of their stony nests like Cadmus' dragon's teeth or the lizards in an M. C. Escher print—all this may seen a bit bizarre. Yet it's really no more strange than imagining the same transformation from droplets of broth. The change happened somehow.

The biggest hurdle for this theory is accepting the idea that life could develop in the dry state, either out of the oceans or in the rocks underneath them. Since the mid-1960s it has seemed more plausible, for it was then that H. E. Hinton, of the University of Bristol, England, learned that at least one organism spends part of its life completely without liquid water. Certain flies of the Sahara desert lay their eggs in the brief pools formed by the rare rains. The larvae go through several metamorphoses in the water, but they're almost always interrupted by the evaporation of the pool. Though completely desiccated, in a state Hinton named cryptobiosis, they survive months or years until the next rainstorm, whereupon they take up where they left off. The larvae can be quick-dried and stored in a vacuum bottle for many years. Placed in water, they resurrect in a few minutes. If a larva is cut in two when active, it takes six minutes to die. If it's flash-dried in the first minute, the two pieces can be kept on a shelf for years, but when returned to water they'll live out their remaining five minutes. Contrary to common sense, it appears that in this case life doesn't need water, but death can't occur without it.

Getting rid of the water-equals-life assumption makes the crystalline theory more believable. Conditions on the young planet favored forests of crystals: It was hot; volcanoes were constantly firing new materials into the dense, dark shell of turbulent gases. However, the crystals would still have needed outside energy to overcome the entropy of nonliving matter. With an organizing principle built into them from the start, it's not too hard to imagine them acquiring other kinds of molecules, including the organics raining from the sky and dissolving in the waters. Then life would have been on its way to developing the biochemistry we now know—the genetic system and the consequent appearance of sexuality—which is the basis of all the creatures now alive or known from the fossil record. Still, we need an energy source for the transition. Lightning won't work in this context. We also need an explanation for the exclusively left- or right-handed molecules.

In 1974 F. E. Cole and E. R. Graf of New Orleans made a theoretical analysis of the Precambrian earth's electromagnetic field that fulfilled both needs. They reasoned that since the atmosphere was much larger then, it must have pushed the ionosphere much farther out than it is today, into the region of the Van Allen belts. The earth would then have had an electromagnetic resonator of two concentric spheres—the upper atmosphere and the surface. Today, as in the past, the earth's pulsing magnetic field combines with the solar wind to induce large currents in the Van Allen belts. In the Precambrian era, however, the fluctuations of current in the Van Allen belts in turn would have generated huge currents in the nearby ionosphere. Since the earth's metallic core is an excellent conductor, the ionospheric currents would have coupled to it, producing an enormous and constant electrical discharge through the atmosphere and into the earth. Moreover, since the distance around the core at that time was roughly equal to 1 wavelength of electromagnetic energy at 10 cycles per second, or about 18,000 miles, this discharge would have pulsed at 10 hertz throughout the resonant cavity, which encompassed the whole atmosphere and surface. Besides directly providing electrical energy, this discharge would have produced abundant heat, ultraviolet radiation, and infrasound (or pressure waves), all of which would have fostered varied chemical activity.

Such a dense and electrically supercharged atmosphere undoubtedly would have produced great quantities of amino acids and peptides. As they came together in the air and water, linking chainwise to form proteins and nucleic acids, the vectors of electromagnetic force would have favored spiral shapes twisting in one direction or the other, depending on whether the reaction occurred in the Northern or the Southern Hemisphere. In 1981, W. Thiermann and U. Jarzak found some direct evidence for this theory by synthesizing organic compounds in a steady-state magnetic field. Changing the orientation of the field gave them high yields of either D or L forms.

It may be possible to run a further check on the Cole and Graf hypothesis at one place in the solar system—the Great Red Spot of Jupiter. This permanent hurricane, whose eye could swallow the earth, continually emits prodigious electrical discharges through an atmosphere much like the one proposed for Precambrian times. It may be synthesizing organic compounds and energizing a transition to life even now.

On earth, all entities formed within the 10-hertz discharge—and all of their descendants—would resonate at the same frequency or show extreme sensitivity to it, even after the original power source had been disconnected. The 10-hertz band would remain supremely important for most life-forms, as indeed it has. As already noted, it's the primary frequency of the EEG in all animals, and it can be used to restore normal circadian rhythms to humans cut off from the normal fields of earth, moon, and sun. William Ross Adey of the Loma Linda VA Hospital in

California has found that magnetic fields modulated at about the same frequency can be used to change the behavior of monkeys in several important ways, which are described more fully in the next chapter.

The Cole and Graf theory also suggests how the spark of life turned itself off. The currents driving the discharges would have ceased as the atmosphere gradually became depleted by escape of the lighter gases and by incorporation of the ammonia and methane into organic compounds. As this happened, the ionosphere would have gradually descended, becoming disconnected from the Van Allen belts. The ionospheric currents would have become too small to couple to the earth's core, and the atmospheric cavity too small to resonate at the core's prescribed frequency. At that point, the plug was pulled, but life was well on its way. Aside from competition from more advanced creatures, the loss of the energy source would explain why we see today no remnants of the transitional forms still emerging from inanimate matter.

This solid-state theory of life's creation is more than an exciting picture of our birth in a shower of sparks. It leads us to another of biology's great mysteries—the evolution of nervous systems—by a sensible sequence of steps. First there would have been a crystalline protocell transmitting information directly through its molecular lattice. As the first cells developed, we can envision chains of microcrystals, then chains of organic polymers transmitting information in the form of semiconducting currents. Although the exact mechanism of electron passage through living tissue is far from clear, nearly all organic matter exhibits piezoelectricity and all the other hallmarks of semiconduction. Furthermore, in a series of experiments during the 1970s, Freeman Cope, a Navy biophysicist building on Szent-Györgyi's work, has found evidence of *super*conduction at room temperature in a variety of living matter. Currents briefly induced in superconductors have been known to flow for many years without decay, but the phenomenon has heretofore been attained only near absolute zero. Although Cope's work is still preliminary and uncorroborated, he has found electromagnetic data consistent with superconduction in *E. coli* bacteria, frog and crayfish nerves, yeast, sea urchin eggs, and molecules of RNA, melanin pigment, and the enzyme lysozyme.

Whatever the exact details of the conducting system, the first multicellular organisms probably had networks of cells that were much like the first single cells. Later, these network cells would have specialized for their DC-carrying duties, linking into syncytia to avoid the high resistance of intercellular junctions. Somewhere along the line a central processing center and information storehouse would have developed. At the same time separate input and output tracts would have appeared, and the DC system would have neared its peak of specialization as its cells evolved into the prototypes of glial, ependymal, and Schwann cells. At about this point the high-speed digital impulse system for handling more complex information would have begun to form inside the older one. Today all multicellular animals have this kind of hybrid system, whose complexities should provide work for at least a few more generations of neurophysiologists.

Crossroads of Evolution

The Cole and Graf theory has one crucial requirement. The polarity of the earth's magnetic field must have stayed the same during the resonant period. Otherwise there would be a mixture of right and left isomers in living tissues. As far as we can tell, the field did remain steady in Precambrian times, but we have ample proof that its poles have reversed many times during the last half-billion years. Each time, the shift has coincided with the extinction of many species.

The geomagnetic record is written in two places: in igneous rocks bearing magnetic minerals, and in ocean floor sediments. Magnetic particles in molten rock are free to move and align themselves with the prevailing magnetic field. As the rock cools they're frozen in place. In the same way magnetic particles settling onto the ocean floor reflect the orientation of the field at the time of the deposit. Ocean sediments and the rock they eventually become have given us an undisturbed magnetic chronology for many millions of years, while the relatively few strata of igneous rock undisturbed by later upheavals give us occasional glimpses further back.

The reversals happen very fast, as geologic time goes. The field strength falls to about half its average for a few thousand years. Then during another thousand years the poles change places; then the field regains its normal strength in another few thousand years. All told, the change takes about five thousand years.

In the early 1960s, when the reversals were first discovered, geophysicists believed the magnetic field disappeared completely during the pole reversal. Thus they thought that the absence of the electromagnetic umbrella that protected life from high-energy ultraviolet and cosmic rays would account for large-scale extinctions. These "great dyings" had long puzzled paleontologists. Soon the demise of a species of radiolarian was correlated with a magnetic field reversal. Radiolarians are microscopic plankton animals with hard calcareous skeletons. Each species has a distinct, intricate shape, so their remains form an easily recognizable, continuous record in sediment cores. By 1967 James D. Hayes and Neil D. Opdyke of Columbia's Lamont-Doherty Geological Observatory had correlated the disappearance of eight types of radiolarians with the reversals. Each species had been widespread and abundant; each extinction took place abruptly, with no previous decline in population. The "radiation barrage" theory seemed confirmed.

However, it has since been learned that the field strength drops only by half, not enough to drastically reduce the protective power of the Van Allen belts and ionosphere. Moreover, radiolarian populations extend down into several yards of water, which should protect them from the radiation anyway. Hays has since drawn the less specific outlines of current knowledge thus: As animals grow more specialized in the course of evolution, they become more sensitive to some as yet unknown, lethal effect of the reversals. Long periods without reversals—the quiescent eras sometimes last tens of millions of years—seem to produce a profusion of species especially susceptible to the effect, and they're weeded out at the next shift.

We know of two especially widespread extinctions. One, at the end of the Permian period, about 225 million years ago, wiped out half the kinds of animals then alive, from protozoa to early reptiles. The same kind of curtain dropped on the age of dinosaurs at the end of the Cretaceous period, some 70 million years ago. In both cases frequent magnetic pole reversals had resumed after a long quiescence. Many periods of less extensive extinction have also been documented in the fossil record and correlated with the field reversals. Most recently, J. John Sepkoski, Jr., and David M. Raup of the University of Chicago reported what they believed to be a 26-million-year cycle in the major dyings. If their hypothesis holds up, there may be some solar or galactic influence that interacts with a magnetic reversal for maximum destructive effect.

We can only surmise that the earth's field was instrumental in life's beginning, but by 1971 we knew virtually for certain that its polarity shifts had shaped life's development by a "pruning" of species. That year I was invited to a private meeting at Lamont to talk about the reversals, the sole M.D. among a score of biologists and geophysicists. At that time we could only speculate as to how the extinction effect came about. We didn't even have a workable theory of what changes inside the earth caused the turnabouts, or how the process affected the micropulsations and other aspects of the field. All we could agree on was that there were probably changes in every aspect of it, and our knowledge hasn't progressed much since then.

The pole shift happens so slowly that living things may well adapt to it easily; the 50-percent decline in field strength also seems rather unimportant. However, since we know the micropulsations control biocycles, including the timing of the mitotic rhythm, a major change in their frequency could be catastrophic. Experiments with artificial extremely low frequency fields (see Chapter 15) have shown that vibrational rates near normal but slightly above, from about 30 to 100 hertz, cause dramatic changes in the cell cycle time. This interferes with normal growth of the embryo and may tend to foster abnormal, malignant growth as well. If a geomagnetic reversal raises the micropulsation frequencies into this range, the accumulation of growth errors over many generations could well mean extinction.

We have no way of making a forecast, however. Reversals seem to happen at widely varied intervals, as often as every fifty thousand years during some periods, many millions of years apart during other times. The last one seems to have occurred about seven hundred thirty thousand years ago. Several scientists have interpreted data from NASA's MAGSAT orbiter, and from measurements of magnetic particles in lake sediments, as indicating that the earth's magnetic field strength is steadily declining, and has been for the last few thousand years. If so, we may already be entering the next reversal, but it's also possible we're merely experiencing one of the field's many short-term variations.

Nor can we be sure how serious a reversal would be for us. Hominids have weathered them in the past, but we have an extra reason for being uneasy this time. If we're entering a reversal now, it will be the first one in which the normal field is contaminated with our own electromagnetic effluvia, and the most powerful of these, at 50 and 60 hertz, fall right in the middle of the "danger band" in which interference with growth controls can be expected.

The field giveth as well as taketh away, however. If we can hang on until the next peak of its strength, we may benefit from a subtle infusion of electromagnetic wisdom. An ingenious theory recently proposed by Francis Ivanhoe, a pharmacologist and anthropologist at two universities in San Francisco, suggests how important it may have been to our own development.

Ivanhoe made a statistical survey of the braincase volume of all known Paleolithic human skulls, and correlated the increase with the magnetic field strength and major advances in human culture during the same period. Ivanhoe found bursts of brain-size evolution at about 380,000 to 340,000 years ago, and again at 55,000 to 30,000 years ago. Both periods corresponded to major ice ages, the Mindel and Würm, respectively, and they were also eras when great cultural advances were made—the widespread domestication of fire by *Homo erectus* in the early Mindel, and the appearance of *Homo sapiens sapiens* (Cro-Magnon peoples) and gradual decline of Neanderthals (*Homo sapiens*) during the Würm. Two other glaciations in the same time span—the Ganz of about 1,200,000 to 1,050,000 years ago and the Riss of about 150,000 to 100,000 years ago—didn't call forth such obvious advancements in human evolution. They also differed from the other two in that the average geomagnetic field intensity was much lower.

Ivanhoe has proposed a direct link from the magnetic field through the growth-hormone regulator pathways in the brain to account for the sharp evolutionary gains. He suggests that part of the hippocampus, a section of the brain's temporal lobe, acts as a transducer of electromagnetic energy. A part of the hippocampus called Ammon's horn, an arch with one-way nerve traffic directed by a strong current flow, may read out variations in the field strength, feeding them by a bundle of well-documented pathways called the fornix to the hypothalamus and thence to the anterior pituitary, where growth hormone is produced. It's known that larger amounts of this hormone in pregnancy increase the size of the cerebral cortex and the number of its nerve cells in the offspring, as compared with other parts of the brain. Ivanhoe also notes that the hippocampus and its connections with the hypothalamus are among the parts of the brain that are much larger in humans than other primates. The idea gains further support from the fact that neural activity in the hippocampus increases with electrical stimulation and reaches a maximum at 10 to 15 cycles per second, at or slightly above the dominant micropulsation frequency of today's field. The most powerful shaper of our development may turn out to be the subtlest, a force that's completely invisible to us.

Hearing Without Ears

We've considered how the electromagnetic fields of earth, moon, and sun affect life. In the next chapter, we'll ponder the effects of artificial fields from our machines. There's probably another interaction, however, of which we know much less: the effects produced on living things by the biomagnetic fields of other creatures. If one nervous system could sense the field of another, it would go a long way toward explaining extrasensory perception.

Following the curious dogma that what we don't understand can't exist, mainstream science has dismissed psychic phenomena as delusions or hoaxes simply because they're rarer than sleep, dreams, memory, growth, pain, or consciousness, which are all inexplicable in traditional terms but are too common to be denied. Fifty years ago, when J. B. Rhine of Duke University first published results of his card-guessing experiments, scientists eagerly debated and tested the subject for a few years. Then, although at least 60 percent of the attempts to confirm Rhine's work also got better-than-chance results (a replication rate better than that in most other areas of psychology), the openness somehow disappeared. Ever since World War II, serious parapsychologists have been hounded out of the forum of science. In the 1950s, for example, *Science* and *Nature* both published attacks on certain results of Rhine and S. G. Soal, an early psi researcher at London University. Today this attitude may be waning. G. R. Price, the author of one of the diatribes, apologized in *Science* in 1972, and both journals have begun accepting occasional reports on psychic research, although still confining themselves mainly to negative findings. As the climate has begun to change, a few researchers have looked for electromagnetic fields as a possible basis for extrasensory perception.

The results so far have been as inconclusive as those from any other approach. In 1978 E. Balanovski and J. G. Taylor used a variety of antennae, skin electrodes, and magnetometers to monitor a number of people claiming paranormal powers. They found no electric or magnetic fields associated with successes in telepathy experiments. In 1982, Robert G. Jahn, dean of engineering at Princeton, assembled the most impressive battery of electronic equipment ever brought to bear on the subject. He found definite effects by mental forces on interferometer displays and strain-gauge readings, along with positive results in remote-viewing experiments. The tests couldn't be reliably repeated, however, and seemed to vary with the moods of researcher and subject, and perhaps with other immeasurable environmental factors. The same dependence on attitude—experiments seem to work more often for believers than doubters—has bedeviled psychic research from its beginning. Although Jahn came up with no clear-cut findings on electromagnetic factors, he was forced to the sublimely understated conclusion that ". . . once the illegitimate research and invalid criticism have been set aside, the remaining accumulated evidence of psychic phenomena comprises an array of experimental observations . . . which compound to a philosophical dilemma."

We must remember that our study of biofields is still in its infancy. It's only a decade since the SQUID first enabled us to find the magnetic field around our heads at all. Pigeons have a magnetic detector thousands of times more sensitive than the latest instruments. We also know that the interaction of semiconducting currents with external magnetic fields is thousands of times greater than that of currents in a wire, and engineers have built microscopic devices that enhance this sensitivity by a factor of another thousand or more. The electron microscope has shown us crystallike structures of previously unsuspected complexity in all living cells, whose functions we can only guess at. There's now some evidence that psychic intent can influence the flow of current in solid-state devices, so we may be nearing the energy levels at which extrasensory factors work. Since all living things generate weak electromagnetic fields, and since many, if not all, can sense those of the earth, communication by this medium remains a strong possibility. Recent disclosure of a multimillion-dollar research effort in this area by the hardheaded weapons planners at the Department of Defense is one more reason why those scientists who work in public shouldn't dismiss the idea.

We must always be careful to place more weight on observation than current theory. We must remember that we don't yet fully understand magnetism. It now appears that the single domain with both magnetic poles may not be the smallest unit of magnetism after all. Physicists now posit the existence of magnetic monopoles, particles having the characteristics of just one pole, north or south. In fact there's some experimental evidence for them. Some theoreticians go even further, envisioning a hitherto unsuspected *kind* of magnetism, a composite of waves and monopole particles, like light. Living things may interact with such a now immeasurable energy.

Any such message system would have at least two major difficulties to overcome in the course of evolution. Our own electrical-engineering experience, however, suggests workable approaches life may have taken.

One problem is that the strength of biofields is far below that of the earth's field. Hence any input from other creatures would be embedded in noise. This is a common obstacle to telecommunications, and there are several ways around it. The easiest is for sender and receiver both to be frequency locked, that is, tuned to one frequency and insensitive to others. Such a lock-in system might explain why spontaneous ESP experiences most often happen between relatives or close friends. The sensitivity of our instruments may someday develop to the point where we can tune in to biomagnetic fields on select frequencies, thus experimenting as directly with ESP as we now do with radio.

Another theoretical difficulty is the fact that psychic transmission doesn't seem to fade with distance. The electromagnetic field around an animal's nervous system, on the other hand, starts out unimaginably small and then diminishes rapidly. However, extremely low frequency (ELF) transmissions have a peculiar property. Because of their interaction with the ionosphere, even weak signals in this frequency range (from 0.1 to 100 cycles per second) travel all the way around the world without dying out. If an innate frequency selector is operating within this band, reception should be the same anywhere on earth.

At this time the DC perineural system and its electromagnetic fields provide the only theory of parapsychology that's amenable to direct experiment. And it yields hypotheses for almost all such phenomena except precognition. Telepathy may be transmission and reception via a biologically programmed channel of ELF vibrations in the perineural system's electromagnetic field. Dowsing may involve an unconscious sense of the electromagnetic fields of underground water or minerals, an idea given some support by Russian experiments in the 1960s. Nikolai N. Sochevanov, now with the USSR's Ministry of Geology, found that the accuracy of forty professional dowsers diminished by at least three fourths when he wound a current-carrying wire around their wrists or brought a horseshoe magnet near their heads.

Biological semiconductors even offer a possible basis for the aura often reported around living things by "sensitives." There has long been speculation that this "halo" might be some manifestation of an electromagnetic biofield. The ability of high-voltage (Kirlian) photography to produce an image very much like descriptions of the aura has aroused hope that the technique might render some aspect of psychic phenomena visible in a way that would be conducive to experiment. Because of this possibility, our lab investigated Kirlian photography during the mid-1970s.

We obtained beautiful pictures that seemed to vary in response to changes in the health of the test organism. However, the method failed one crucial test. If the Kirlian halo actually reflected the biofield or some other basic aspect of life, it should have disappeared when the organism being photographed died. Alas, it did not. The image remained the same as long as the water content of the corpse remained constant. We found the images were entirely due to a simple physical event, a corona discharge. This occurred when a high-voltage electric field broke down the air molecules between the two condenser plates of the Kirlian apparatus. The amount of water vapor in the air changed the voltage at which this happened, and on color film produced coronas in different colors and sizes. We found no evidence that the Kirlian image was related to the living state. Nor did we find that it could serve as a "screen" on which might be reflected some invisible field or aura, another possibility that had been suggested.

This is not to say that the aura occasionally perceived by some people around other organisms is imaginary. Things that appear so often in folklore often turn out to have a basis in fact. However, the body's magnetic field is far too weak to account for it. Our biofields, even if they were many times stronger, couldn't possibly emit light, but an appropriately sensitive magnetic detector in the brain, if it had nerve connections to the visual cortex, might "see" the magnetic field, in a manner of speaking. In a similar way astronauts in space "see" Cerenkov radiation—flashes of light that have been traced to the passage of high-energy cosmic rays through the retina.

On the other hand, the aura could literally be a form of light, perhaps at frequencies invisible to all but a few of us. The discovery of light-emitting diodes is still fairly recent. As you will recall, we found that bone happens to have such properties. The point of that experiment was its evidence that bone contains semiconducting PN junction diodes. There may well be other diodes in living things. The relationship between the nerve endings and the skin is an interesting one in this context. The skin-nerve interface—the closest normal equivalent to the neuroepidermal junction that triggers regeneration—may well be a diode. If so, the proper level of current could cause emission of light from the skin. It's even possible that such an array of diodes with very large currents might produce a holographic image of the body on an organic screen, such as the reputed image of Christ on the Shroud of Turin.

If extrasensory communication really is a function of the DC system, why isn't it more common and widely accepted? We may never know how well distributed it is among animals, although the number of pets who have returned to their owners over long distances suggests that many dogs and cats can find specific people by an unknown sense. The Duke University Parapsychology Laboratory has authenticated more than fifty such cases, many involving travel of hundreds or thousands of miles. We can expect that some species would be better at it than others, just as pigeons navigate by the earth's magnetic field far better than most other creatures. Among humans, some may simply be more gifted than others through genetic chance or some facet of their upbringing. Then again, the psychic sense may be a universal ability that was forgotten or suppressed as we came to depend more and more on language to get our messages across.

If they do depend on the same system, psychic ability and regeneration may go together; they may generally be better among simple animals. As the digital impulse system grew more efficient, its information may have overwhelmed the senses operating through the earlier mode. In fact, this may be part of the digital system's purpose. The ever-present hum of electromagnetic information from other creatures may have become an intolerable burden. Think how confused you would feel if you could simultaneously hear what everyone else in the world was thinking. After all, mediums, sorcerers, and psi experimenters all agree that some sort of trance or mental quietude—a reduction of nerve impulse activity—is needed for best results. According to Elmer Green, yogis of some Tibetan traditions teach clairvoyance to novices by having them meditate seated on a glass plate, facing north toward a sheet of polished copper in a dark, windowless room, with a bar magnet suspended over their heads, its north pole pointing up to the zenith.

The biofield also lends itself to theories of psychokinesis and object imprinting. All matter, living and nonliving, is ultimately an electromagnetic phenomenon. The material world, at least as far as physics has penetrated, is an atomic structure held together by electromagnetic forces. If some people can *detect* fields from other organisms, why shouldn't some people be able to *affect* other beings by means of their linked fields? Since the cellular functions of our bodies are controlled by our own DC fields, there's reason to believe that gifted healers generate supportive electromagnetic effects, which they convey to their patients or manipulate to change the sufferer's internal currents *directly,* without limiting themselves to the placebo effect of trust and hope.

Once we admit the idea of this kind of influence, then the same kind of willed action of biofields on the electromagnetic structure of *in*animate matter becomes a possibility. This encompasses all forms of psychokinesis, from metal-bending experiments in which trickery has been excluded to more rigidly controlled tests with interferometers, strain gauges, and random number generators. At present, it's the only hypothesis that offers much hope of testability. On a less spectacular level, we must ask whether the biofield can project the individual signature of a person's thoughts onto his or her surroundings, changing the electromagnetic characteristics of these objects so that the person can be sensed by others even though absent. This may well be the commonest of all paranormal experiences, and the number of crimes solved by psychics reacting to the mere scene of the crime should entitle scientists to investigate the idea without fear of ridicule from their colleagues.

Over and over again biology has found that the whole is more than the sum of its parts. We should expect that the same is true of bioelectromagnetic fields. All life on earth can be considered a unit, a glaze of sentience spread thinly over the crust. *In toto,* its field would be a hollow, invisible sphere inscribed with a tracery of all the thoughts and emotions of all creatures. The Jesuit priest and paleontologist-phi-

Fifteen

Maxwell's Silver Hammer

In considering questions as remote as the origin of life, science must skate toward new shores across the thin ice of speculation, but it also has a duty to warn us of present dangers as specifically as possible. Since the earth's electromagnetic activity has such a profound effect on life, the obvious question is: What are the consequences of our artificial energies?

Electromagnetism can be discussed in two ways—in terms of fields and in terms of radiation. A field is "something" that exists in space around an object that produces it. We know there's a field around a permanent magnet because it can make an iron particle jump through space to the magnet. Obviously there's an invisible entity that exerts a force on the iron, but as to just what it consists of—don't ask! No one knows. A different but analogous something—an electric field—extends outward from electrically charged objects.

Both electric and magnetic fields are static, unvarying. When the factor of time is introduced, by varying the intensity of the field as in a radio antenna, an electromagnetic field results. As its name implies, this consists of an electric field and a magnetic field. The fluctuations in the field radiate outward from the transmitter as waves of energy, although somehow these waves simultaneously manage to behave as streams of massless, chargeless particles (photons). As to just how this happens, again—don't ask! Sometimes the phenomenon is called an electromagnetic field (EMF), to emphasize its connection with the transmitter; sometimes it's called electromagnetic radiation (EMR), to emphasize its outward-flowing aspect. However, the two terms refer to the same phenomenon and are interchangeable. The only meaningful distinction is between static and time-varying fields.

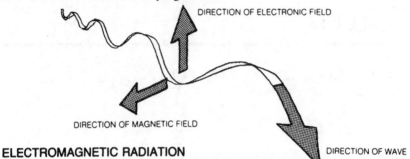

DIRECTION OF ELECTRONIC FIELD

DIRECTION OF MAGNETIC FIELD

DIRECTION OF WAVE

ELECTROMAGNETIC RADIATION

Each energy wave consists of an electric field and a magnetic field at right angles to each other, and both at right angles to the direction the wave is traveling. The number of waves formed in one second is the frequency; the distance the energy travels (at the speed of light) during one oscillation is its wavelength. The higher the frequency, the shorter the wavelength, and vice versa.

EMR spans an enormous range of frequencies. The shortest gamma rays, a tenth of a billionth of a millimeter long, vibrate sextillions of times a second. These, along with X rays and the shortest ultraviolet wavelengths, are termed ionizing radiation, because their high photonic energy can knock electrons away from atoms, creating highly reactive ions where they don't belong. Much of the damage from nuclear radiation is caused in this way. All lower frequencies, beginning with the longer ultraviolet wavelengths, are nonionizing.

Next comes the only energy we can see—the narrow band of visible light vibrating hundreds of trillions of times a second—and then the infrared waves we feel as radiant heat. Below these lie the waves we've harnessed for communication. They begin with microwaves (MW), whose frequency is measured in gigahertz or megahertz—billions or millions of cycles per second—and extend through the radio frequencies (RF) down to ELF waves, whose frequency converges on zero. The MW and RF spectrum is arbitrarily broken up into a further alphabet of extremely high, superhigh, ultrahigh, very high, high, medium, low, very low, and extremely low frequencies (EHF, SHF, UHF, VHF, HF, MF, LF, VLF, and ELF respectively). As we have seen, ELF waves approximate the dimensions of the earth; at 10 hertz one wave is about 18,600 miles long.

Except for light and infrared heat, we can't perceive any of these energies without instruments, so most people don't realize how drastically and abruptly we've changed the electromagnetic environment in just one century. Working at Cambridge University, Scottish physicist James Clerk Maxwell showed mathematically in 1873 that light was but a small part of the vast undiscovered realm of radiation. Heinrich Hertz first found some of the radio waves in 1888. Meanwhile, Edison had set up the first commercial electric-power system in New York in 1882.

For billions of years before then, the energies that life grew up among were relatively simple. There was a weak electromagnetic field modulated by micropulsations within it and further sculpted by the solar and lunar cycles. There was a burst of static centered at 10,000 hertz and reverberating over the whole earth whenever lightning flashed in the scores of thunderstorms in progress at any one time. There were a few weak radio waves from the sun and other stars. Light, including some infrared and ultraviolet, was the most abundant form of electromagnetic energy. At higher frequencies, living things absorbed only small amounts of ionizing X rays and gamma rays from space and from radioactive minerals in rocks. Large parts of the energy spectrum were totally silent.

We'll never experience that quiet world again. In 1893 Nikola Tesla lit the Chicago World's Fair with the first AC power system, and two years later he began the modern era of electrical engineering by harnessing Niagara Falls. In 1901 Guglielmo Marconi sent a radiotelegraph message across the Atlantic, using without acknowledgment a machine designed by the prolific Tesla. The invention of the vacuum tube in 1907 led to the first voice transmission by radio in 1915 and the first commercial station in 1920. Until then many people still ate supper by candle or kerosene, and the ambient forces remained a reasonable facsimile of earth's pristine field.

The greatest changes have all come in the one generation since World War II. The trend toward use of shorter and shorter radio waves, bounced off the ionosphere for long-distance communication, had begun before the war. The fight for survival against fascism impelled the development of microwave radar, which helped win the Battle of Britain, allowed all-weather bombing of Germany, and gave the American Navy a decisive edge over the Japanese. The conflict also produced other electronic devices of all types. In 1947 Bell Telephone set up the first microwave phone relay towers between New York and Boston, the same year the first commercial television broadcasts, also transmitted by microwaves, began. Since then nearly every human action has involved an electrical appliance, and today we're all awash in a sea of energies life has never before experienced, of which the following list of sources only skims the surface:

- Everything that runs on a battery produces a DC magnetic field—from digital watches, cameras, flashlights, and portable radios to car ignition systems.
- Strong magnetic fields are used in industry to refine ore, concentrate and recycle scrap iron, purify sewage, soften water for steam boilers, and many other tasks.
- The starting and stopping of an electric train turns the power rail into a giant antenna that radiates ELF waves for over 100 miles.
- Electromagnetic fields vibrating at 60 hertz (50 hertz in Europe and Russia) surround nearly every person on earth from appliances at home and machines at work.
- Over 500,000 miles of high-voltage power lines crisscross the United States. Innumerable smaller lines feed into every home, office, factory, and military base, all producing AC or DC fields. Metal objects near the lines concentrate the fields to higher levels. In addition, high-voltage lines are, in effect, gigantic antennae operating at 60 hertz in the ELF band, the largest "radio" transmitters in the world. Switching stations, where the current is changed from one voltage or type to another, emit radio-frequency waves as well.
- AC magnetic fields vibrating at 100 to 10,000 hertz emanate from antitheft systems in stores and libraries, and from metal detectors in airports.
- Low-frequency radio waves are used for air and sea navigation, time references, emergency signals, some amateur radio channels, and military communications.
- Medium frequencies between 535 and 1,604 kilohertz are reserved for AM radio transmitters, which are limited to 50,000 watts in this country but are sometimes much more powerful abroad.
- HF and VHF channels are filled with chatter from the nation's 35 million CB radios, as well as shortwave bands for more ham radios, air and sea navigation systems, military uses, spy satellites, and police and taxi radios. VHF television and FM radio also inhabit this region. There are now over ten thousand commercial radio and TV stations in the United States alone, and 7 million other radio transmitters, not counting the millions operated by the military.
- Weather satellites, some kinds of radar, diathermy machines, upward of 10 million microwave ovens, more cop and cab radios, automatic garage-door openers, highway emergency call boxes, and UHF television compete for the low microwave frequencies.

- Higher microwave bands are crowded with more military talk channels and radar, navigational beacons, commercial communications satellites, various kinds of walkie-talkies, and America's two hundred fifty thousand microwave phone and TV relay towers.
- Like the infrared rays above them in the spectrum, radio waves and microwaves produce heat when directed in high-intensity beams. Hence they're used for all sorts of industrial chores—bonding plywood, vulcanizing rubber, manufacturing shoes, sterilizing food, making plastics, and heat sealing the trillions of plastic-wrapped products in our stores, even opening oysters. Modern electronics would be impossible without the perfect silicon and germanium crystals grown in microwave furnaces.

The human species has changed its electromagnetic background more than any other aspect of the environment. For example, the density of radio waves around us is now 100 million or 200 million times the natural level reaching us from the sun. Nor is there any end in sight. When superconducting cables are introduced, they'll increase the field strength around power lines by a factor of ten or twenty. Electric cars, magnetically levitated transport vehicles, and microwave-beam satellites for transmitting solar power to earth would each add strong new sources of electromagnetic contamination. A proposed electromagnetic catapult that could shoot satellites into space from mile-long rails built up the side of a mountain would require the combined output of the country's thousand generating stations for the few seconds of each launch.

A few years ago most investigators believed that each wavelength interacted mainly with objects comparable to it in size. This was a comforting notion that theoretically limited each frequency to one type of effect and predicted that really troublesome problems for humans would come from only one portion of the spectrum—the FM band. Now, however, we know there are primary effects on all life-forms at ELF frequencies, and in other parts of the spectrum there can be consequences for specific systems at *any* level, from the subatomic to the entire biosphere as a unit.

Of course, a change at one level may well trigger secondary changes throughout an organism, so that the original one is hard to identify. Moreover, the impact of EMR at any particular frequency is often related to its power density, the amount of energy streaming through a certain area. When discussing biological effects, this is best measured in microwatts (millionths of a watt) per square centimeter, a unit we'll simplify to microwatts. There's often no direct relationship between dose and effect, however; a low power density sometimes does things that a higher one does not. Furthermore, we can't tell how much energy from a given power density is actually absorbed, or what part of the body receives it. The same holds true for electric and magnetic fields, whose study is further complicated by the fact that animals of different shapes distort the fields differently. Likewise, fur, feathers, skin thickness, bone size, and the general shape of an animal complicate RF and MW absorption beyond our capacity to gauge it. Therefore, reactions seen in one species cannot be assumed for another. The only way to test for possible damage (or beneficial effects) is to actually do the experiment.

In a sense, the entire population of the world is willy-nilly the subject of a giant experiment. Electropollution has been the subject of heated *public* debate for nearly ten years, and unpublicized misgivings for decades before that. Unfortunately, the question of risk has been asked too late. Daily exposure of nearly everyone is a *fait accompli*.

Subliminal Stress

After Howard Friedman, Charlie Bachman, and I had found evidence that "abnormal natural" fields from solar magnetic storms were affecting the human mind as reflected in psychiatric hospital admissions, we decided the time had come for direct experiments with people. We exposed volunteers to magnetic fields placed so the lines of force passed through the brain from ear to ear, cutting across the brainstem-frontal current. The fields were 5 to 11 gauss, not much compared with the 3,000 gauss needed to put a salamander to sleep, but ten to twenty times earth's background and well above the level of most magnetic storms. We measured their influence on a standard test of reaction time—having subjects press a button as fast as possible in response to a red light. Steady fields produced no effect, but when we modulated the field with a slow pulse of a cycle every five seconds (one of the delta-wave frequencies we'd observed in salamander brains during a change from one level of consciousness to another), people's reactions slowed down. We found no changes in the EEG or the front-to-back voltage from fields up to 100 gauss, but these indicators reflect major alterations in awareness, so we didn't really expect them to shift.

We were excited, eagerly planning experiments that would tell us more, when we came upon a frightening Russian report. Yuri Kholodov had administered steady magnetic fields of 100 and 200 gauss to rabbits and found areas of cell death in their brains during autopsy. Although his fields were ten times as strong as ours, we stopped all human experiments immediately.

Friedman decided to duplicate Kholodov's experiment with a more detailed analysis of the brain tissue. He made the slides and sent them to an expert on rabbit brain diseases, but coded them so no one knew which were which until later.

The report showed that all the animals had been infected with a brain parasite that was peculiar to rabbits and common throughout the world. However, in half the animals the protozoa had been under control by the immune system, whereas in the other half they'd routed the defenders and destroyed parts of the brain. The expert suggested that we must have done something to undermine resistance of the rabbits in the experimental group. The code confirmed that most of the brain damage had occurred in animals subjected to the magnetic fields. Later, Friedman did biochemical tests on another series of rabbits and found that the fields were causing a generalized stress reaction marked by large amounts of cortisone in the bloodstream. This is the response called forth by a prolonged stress, like a disease, that isn't an immediate threat to life, as opposed to the fight-or-flight response generated by adrenaline.

Soon thereafter, Friedman measured cortisone levels in monkeys exposed to a 200-gauss magnetic field for four hours a day. They showed the stress response for six days, but it then subsided, suggesting adaptation to the field. Such seeming tolerance of continued stress is illusory, however. In his pioneering lifework on stress, Dr. Hans Selye has clearly drawn the invariable pattern: Initially, the stress activates the hormonal and/or immune systems to a higher-than-normal level, enabling the animal to escape danger or combat disease. If the stress continues, hormone levels and immune reactivity gradually decline to normal. If you stop your experiment at this point, you're *apparently* justified in saying, "The animal has adapted; the stress is doing it no harm." Nevertheless, if the stressful condition persists, hormone and immune levels decline further, well below normal. In medical terms, stress *decompensation* has set in, and the animal is now more susceptible to other stressors, including malignant growth and infectious diseases.

In the mid-1970s, two Russian groups found stress hormones released in rats exposed to microwaves, even if they were irradiated only briefly by minute amounts of energy. Other Eastern European work found the same reaction to 50-hertz electric fields. Several Russian and Polish groups have since established that after prolonged exposure the activation of the stress system changes to a depression of it in the familiar pattern, indicating exhaustion of the adrenal cortex. There has even been one report of hemorrhage and cell damage in the adrenal cortex from a month's exposure to a 50-hertz, 130-gauss magnetic field.

Soviet biophysicist N. A. Udintsev has systematically studied the

effects of one ELF magnetic field (200 gauss at 50 hertz) on the endocrine system. In addition to the "slow" stress response we've been discussing, he found activation of the "fast" fight-or-flight hormones
centering on adrenaline from the adrenal medulla. This response was
triggered in rats by just *one day* in Udintsev's field, and hormone levels
didn't return to normal for one or two weeks. Udintsev also documented
an insulin insufficiency and rise in blood sugar from the same field.

One aspect of the syndrome was very puzzling. When undergoing
these hormonal changes, an animal would normally be aware that its
body was under attack, yet, as far as we could tell, the rabbits were not.
They showed no outward signs of fear, agitation, or illness. Most humans certainly wouldn't be able to detect a 100-gauss magnetic field, at
least not consciously. Only several years after Friedman's work did anyone find out how this was happening.

In 1976 a group under J. J. Noval at the Naval Aerospace Medical
Research Laboratory at Pensacola, Florida, found the slow stress response
in rats from very weak electric fields, as low as five thousandths of a volt
per centimeter. They discovered that when such fields vibrated in the
ELF range, they increased levels of the neurotransmitter acetylcholine in
the brainstem, apparently in a way that activated a distress signal sub-
liminally, without the animal's becoming aware of it. The scariest part
was that the fields Noval used were well within the background levels of
a typical office, with its overhead lighting, typewriters, computers, and
other equipment. Workers in such an environment are exposed to electric fields between a hundredth and a tenth of a volt per centimeter and
magnetic fields between a hundredth and a tenth of a gauss.

Power Versus People

Because industry and the military demand unrestricted use of electromagnetic fields and radiation, their intrinsic hazards are often
compounded by secrecy and deceit. I learned this lesson in my first encounter with the environmental review process.

As we were investigating the EMF-stress connection in 1969, the
Navy decided to build a giant antenna in northern Wisconsin. The plan,
called Project Sanguine, was to establish a radio link with nuclear submarines at their normal depth of 120 feet or below. Conventional radio
signals couldn't pass through water, so the vessels had to surface or else
cruise very slowly a few feet under and communicate by means of a
floating antenna at prearranged times. Since this made the subs tem-

porarily vulnerable, the Navy wanted a message system using ELF waves, which penetrate earth and water.

The original design involved 6,000 miles of buried cable arranged in a grid across the upper two fifths of Wisconsin. A transmitter would pump current into one side; the electricity would emerge from the other side and complete the circuit by traveling through the ground. The device was actually a giant loop antenna using the earth as part of the loop. ELF waves issuing from it and resonating between the earth's surface and the ionosphere could be picked up anywhere on the globe.

Sanguine was one of the first military projects scrutinized under the Environmental Protection Act. In 1973 the Navy set up a committee of scientists to review fifteen years of naval research on ELF effects, as well as other pertinent work. Captain Paul Tyler of the Office of Naval Research asked me to be one of its seven members.

The only thing sanguine we found was the name. While the research to date didn't *prove* there would be grave harm to human health, it showed several dangers. The antenna would produce an electromagnetic field 1 million times weaker than that from a 765-kilovolt power line. It was to broadcast at 45 to 70 hertz, frequencies close enough to the earth's micropulsations that living things are very sensitive to them. Similar fields had been shown to raise human blood triglyceride levels (often a harbinger of stroke, heart attack, or arteriosclerosis), and change blood pressure and brain wave patterns in experimental animals. The generalized stress response, desynchronized biocycles, and interference with cellular metabolism and growth processes—and hence increased cancer rates—were also distinct possibilities. Hundreds of thousands of people would be living *inside* the antenna even in this sparsely populated area; long-term effects on plants and animals were unknown; and, because the signals would resonate throughout the world, the biohazards might be similarly widespread. For these reasons we unanimously recommended that the project be shelved pending answers to the ominous questions it raised. We provided a long list of necessary research, emphasizing further tests on triglycerides, biorhythms, stress, and psychological responses to ELF fields. We also warned that the health of a large part of the U.S. population might already be impaired by 60 hertz power lines carrying vastly more power than the proposed antenna.

The committee met on December 6 and 7, 1973, generating a report then and there, with a secretary taking down our conclusions. The Navy group in charge was apparently displeased with our findings. The printed proceedings, marked "For official use only," went out only to committee members, and the Navy refused to discuss them with anyone else.

As soon as I got back from Washington, I found that two power companies were planning a network of 765-kilovolt power lines linking nuclear reactors in upstate New York and Canada. One of the lines was to pass through a rural area near the village of Lowville, where I'd just bought land for a vacation-retirement home. I immediately wrote the head of the state's Public Service Commission. Without releasing the Sanguine report—I felt it wasn't my place to do so, even though its suppression was wrong—I informed PSC Chairman Alfred Kahn of its major conclusions. The commission in turn asked the Navy for a copy of our report but was turned down flat. In mid-1974, however, Andy Marino and I were asked to testify at PSC hearings on the power lines.

We presented the best evidence then available, some of which seemed to shock the PSC members. ELF fields at power line intensity or less had by then been linked to bone tumors in mice, slowed heartbeat in fish, and various chemical changes in the brain, blood, and liver of rats. Bees exposed to a strong ELF field for a few days in Russian research had begun to sting each other to death or leave the area. Some sealed off their hives and asphyxiated themselves. Attorneys for the power companies hurriedly asked a year's postponement of the hearings, which the PSC naturally granted.

Andy and I spent that year reading the rapidly accumulating scientific literature on EMF biological effects, including the enormous amount of Russian work becoming available in English. Andy also investigated the stress response further. He ran ten separate experiments with rats, exposing them for one month to 60-hertz electric fields of 100 to 150 volts per centimeter, simulating ground level underneath a typical high-tension line. Three generations of rats bred in this field showed severely stunted growth, especially among males. At lower field strengths (35 volts per centimeter) some of the animals gained *more* weight than controls, a response we tentatively traced to abnormal water retention, which, like underweight, could also result from stress. A few years later, a study commissioned by the Department of Energy to duplicate this research also produced contradictory but disquieting results. With every known variable controlled in an expensive, high-tech facility at Battelle Laboratories in Columbus, Ohio, one test showed severe growth retardation over three generations, while a second run under exactly the same conditions produced significantly greater weight gain than normal.

Andy's original work also revealed large increases in the infant mortality rate. Between 6 and 16 percent of the pups born in various tests failed to live to maturity because of the electric field. That is, these percentages were in *excess* of the normal death rate for newborn rats.

Various other symptoms consistent with stress were found, including decreased water intake, enlarged adrenal and pituitary glands, and altered protein and hormone ratios in the blood. There was also a very high incidence—ten in sixty—of glaucoma in the early experiments. The disease didn't show up in later runs from which we excluded animals having observable eye defects, suggesting that the electric field had worsened a preexisting problem rather than causing it.

We expected the utilities to roll out their heavy artillery when the PSC hearings resumed, but we were still unprepared for what actually happened. The companies had hired two microwave researchers, Herman Schwan and Solomon Michaelson, both of whom did most of their work for the Department of Defense, and University of Rochester botanist Mort Miller. Carefully prepared by these three, the company lawyers cross-examined us for seventeen days in December 1975, attacking not only our methods and results but our scientific competence and honesty as well. Michaelson strenuously denied that our rodents had shown signs of stress, even though the biological markers were clear. Even if they had, he contended, stress could be healthful, an idea that Hans Selye later called "farfetched" when applied to a biological challenge that was continuous and not self-imposed.

As far as I know, our testimony was the first ever openly given by American scientists stating that electromagnetic energy had health effects in doses below those needed to heat tissue, and that power lines might therefore be hazardous to human health. We criticized the White House Office of Telecommunications Policy for failing to follow up a tentative 1971 warning by advising the President that some harmful effects from electropollution were now proven. Moreover, although we didn't realize it at the time, we greatly embarrassed Captain Tyler and the Navy by publicly revealing the existence of the Sanguine report, which had been secret until then.

Among those who heard about it was Wisconsin Senator Gaylord Nelson, who was understandably furious that his constituents were even then being used as guinea pigs in ongoing ELF tests at an experimental station near Clam Lake, while the document gathered dust in some Navy safe. Quoting Andy and me, he soundly criticized the Navy on the Senate floor. Due to local opposition, the Navy had already moved the site of the full-scale antenna to Michigan's upper peninsula, modifying the design and giving it a new name—Project Seafarer. Nelson's fury now induced the Secretary of the Navy to ask the National Academy of Sciences for further study of the environmental questions. Harvard's biology chairman, Woodland Hastings, who was picked to head the NAS committee, wrote Marino a flattering letter asking for his consultation when the members got down to work. Marino then called Hastings to tell him of the large body of data we'd assembled for the power line hearings, and to make sure the NAS body would be willing to consider it thoroughly. Hastings told him, "Heck, you guys will be *on* the committee."

Soon the sixteen members were announced, and we were nowhere in sight. Hastings later publicly called us quacks, but to us he said the Navy had specified who was to be on the panel, despite his threats to quit if Andy and I weren't admitted. We were well acquainted with three of the men who *were* on it: Schwan, Michaelson, and Miller. Obviously they weren't about to find hazards in Seafarer after testifying that a much stronger power line was perfectly safe. They remained on the committee even though all three neglected to mention their New York testimony on NAS conflict-of-interest forms. The rest of the committee was also stacked with people who routinely discounted any evidence of health effects from low-level EMFs.

The NAS committee took an inordinately long time to issue its report, but we eventually saw a reason for the delay. During the PSC hearings, all evidence was subject to cross-examination. Besides questioning the witnesses, each side could look at the other's papers, including the actual workbooks of experiments. After the testimony, while the commission, assisted by a panel of judges, was deliberating, other evidence could be introduced, but it was no longer subject to review by the opposing side. Oddly enough, the NAS report—which constituted a defense of the then current dogma and tried to discredit most of the disturbing evidence—appeared just after the gavel sounded to close the PSC hearings. It was immediately introduced as evidence, and we couldn't say a word about it.

Six years later a Navy spokesman explained to me what had "really" happened. He said the Secretary of the Navy had gone to NAS and arranged to pay for the work. Then, when the members of the committee were announced, the Secretary and other Navy brass agreed that the show was rigged. The Secretary protested to NAS and said the Navy wouldn't pay for the study. NAS said that since the authorizations had already been signed, the Navy would have to pay for it. Moreover, the Navy needed a report in four to six months. Of course, NAS had been planning to wait till the end of the New York PSC hearings, which dragged on and on. My informant told me that, in response to Navy pressure, NAS said in effect, "Go away. We've got the money, and the study is out of your hands. We'll run it our way." By the time the report was issued it was too late for the Navy people to use it, and they considered it too biased to have any value anyway. However, I don't put much faith in this bit of blame shifting.

The PSC's panel of judges spent nearly a third of their advisory opinion attacking Marino's work and his "argumentative" demeanor at the witness table. Via a Freedom of Information Act request, Andy later found that the technical parts of this opinion had been written by one of the judges' paid consultants, Asher Sheppard, then a researcher at UCLA. Sheppard was at that time preparing a monograph, *The Biological Effects of Electric and Magnetic Fields of Extremely Low Frequency,* under contract to the American Electric Power Company. He concluded that there were no significant biological effects from low to moderate-intensity ELF fields such as occurred around power lines and appliances, despite the fact that he'd been working under W. Ross Adey, whose career has been devoted to studying just such effects.

Nevertheless, we won. The Public Service Commission specifically contradicted its judicial advisers, commending Marino as a valuable witness, and adopted most of our recommendations. One line already under construction was built, largely because New York Governor Hugh Carey threatened to dissolve the PSC if the commission stopped it, but the utilities were ordered to buy additional land for a wider safety zone along the right of way. They were also forced to invest $5 million in a five-year research program administered by the New York State Department of Health, and to stop encouraging multiple use of the land under power lines, such as leasing it for playgrounds. An additional six or seven proposed lines have been postponed indefinitely. Most important was the plain fact that we raised the issue successfully against great odds and secured a health-conscious verdict from the PSC, gaining time to gather more facts about the dangers.

The Navy's ELF antenna has also been on hold for many years. Seafarer lost momentum when 80-percent opposition in two 1976 referenda in Michigan's Upper Peninsula forced then candidate Jimmy Carter to oppose it publicly for a while. Once more renamed and redesigned, Project ELF has been heavily funded by the Reagan regime with an eye toward expansion. The first step now would consist of a 56-mile aboveground antenna carried on intersecting rows of utility poles in two corridors cut out of the Escanaba River State Forest. In July 1983, the Michigan Natural Resources Commission voted to allow construction. However, six months later a federal district judge upheld the suit of several local groups, on whose behalf I testified, ruling that the Navy must prepare a new environmental impact statement. The Navy lost two appeals of this decision but won a lifting of the injunction in the third appeal, so construction is, at this writing, going on.

Fatal Locations

Subliminal activation of the stress response is one of the most important effects that EMFs and nonionizing radiation have upon life, but it's far from the only one. These unfamiliar energies produce changes in nearly every bodily function so far studied. Many of these alterations are associated with stress, but whether they're a result of, or an additional trigger for, the adrenocortical reaction is an irrelevant chicken-or-egg question at this stage of our knowledge. The most disturbing data come from work on the systems that integrate other bodily functions—the central nervous, cardiovascular, endocrine, and growth control systems. We'll concentrate upon these in the following overview of the biohazards.

For the most part, no attempt will be made to identify specific effects from microwaves, radio waves, and electric or magnetic fields, for similar changes have been observed from all modalities. The major problems come from extremely low frequencies, but higher frequencies have the same effects if pulsed or modulated in the ELF range. This is very often the case, for, to transmit information, microwaves or radio waves must be shaped. This is done by interrupting the beam to form pulses or by modulating the frequency or amplitude (size) of the waves. Furthermore, today's environment is a latticework of crisscrossing signals in which there's always the possibility of synergistic effects or the "construction" of new ELF signals from the patterns of interference between two higher frequencies. Therefore, experiments in which cells or organisms are exposed to a single unmodulated frequency, though sometimes useful, are irrelevant outside the lab. They're most often done by researchers whose only goal is to be able to say, "See, there's no cause for alarm."

The Central Nervous System

Since our work on human reaction time, half a dozen other groups have also found marked CNS effects from ELF fields. Most experiments have shown a decrease in reaction speed, although one researcher noted faster-than-normal reactions in humans exposed to very weak electric fields vibrating at beta wave frequencies. The sensitivity of some animals has turned out to be amazing. James R. Hamer of Ross Adey's group at UCLA reported changes in monkeys' response times from ELF electric fields as weak as 0.0035 volts per centimeter, roughly equivalent to the field from a color TV set 60 feet away.

One of the most telling tests was a simple one done at the Navy's Pensacola lab. R. S. Gibson and W. F. Moroney measured people's short-term memory and their ability to add sets of five 2-digit numbers in the presence of a 1-gauss magnetic field—the strength found near some high-voltage power lines and many common high-current appliances, such as portable electric heaters. Test scores declined at both the 60-hertz power frequency and the 45-hertz frequency of the Sanguine-Seafarer antenna, but remained normal in control sessions.

Several studies on both sides of the Iron Curtain have found that rats are generally less active and less exploratory of their environment after being dosed with microwaves, although some frequencies induce restlessness. In contrast, ELF magnetic or electric fields almost always produce hyperactivity and disturbed sleep patterns in rats.

Obviously the subtle workings of the mind may undergo many shifts that don't show up in these crude behavioral tests. Most of our knowledge of electropollution's effects on the brain concerns variables that can be more easily quantified, such as changes in biochemistry, cells, and EEG patterns. These studies can't be easily related to changes in thought processes, but most of the results fit in well with the stress response.

In 1966, Yuri Kholodov found effects on rabbits' EEGs from a few minutes' exposure to fairly strong steady-state magnetic fields (200 to 1,000 gauss). As we'd found in salamanders, there were more delta waves, as well as bursts of alpha waves. He and another Russian biophysicist, R. A. Chizhenkova, also noted a desynchronization, or abrupt shift in the main EEG rhythm, for a few seconds when any field was switched on or off. The same effect has since been confirmed in rats with microwaves. This proved that the brain could sense the field, whether the animal knew it or not.

The sites of the greatest changes—the brain's hypothalamus and cortex—were cause for concern. The hypothalamus, a nexus of fibers linking the emotional centers, the pituitary gland, the pleasure center, and the autonomic nervous system, is the single most important part of the brain for homeostasis and is a crucial link in the stress response. Any interference with cortical activity, of course, would disrupt logical and associational thought.

In 1973 Zinaida V. Gordon, a pioneer in microwave research working with M. S. Tolgskaya at the USSR Academy of Medical Sciences Institute of Labor Hygiene and Occupational Diseases, reported a possible cellular feature of EMR stress. Low doses of microwaves, a mere 60 to

320 microwatts for an hour a day, changed nerve cells in the hypothalami of rats. During the first month of exposure, the neurotransmitter-secreting portions of the cells connecting the brain to the pituitary gland were enlarged. After five months they'd begun to atrophy. When microwave dosage was stopped at that point, however, the cells recovered. J. J. Noval's finding that ELF electric fields changed brainstem acetylcholine levels has already been mentioned. In similar experiments, others have noted a rise followed by a drop to below normal in rat brain levels of norepinephrine, the main neurotransmitter of the hypothalamus and autonomic nervous system. In Soviet work, microwave densities of 500 microwatts or more, delivered in a work-exposure pattern of seven hours a day, gradually reduced norepinephrine and dopamine (another neurotransmitter) to brain levels that indicated exhaustion of the adrenal cortex and autonomic system.

Two years after the Gordon-Tolgskaya report, Allen Frey, who has studied bioeffects of microwaves for over two decades at Randomline, Inc., a consulting firm in Huntingdon Valley, Pennsylvania, found an effect on the blood-brain barrier, the cellular gateway by which specialized capillaries strictly limit the molecules admitted to the delicate nerve cells' environment. Even at power densities as low as 30 microwatts, microwaves pulsed at extremely low frequencies loosen this control, in effect opening up leaks in the barrier. Since some barrier changes occur in response to stress and mood shifts, this could be either a cause or a result of the stress response, or an unrelated effect of pulsed microwaves. In any case, since the blood-brain barrier is the central nervous system's last and most crucial defense against toxins, we must consider this increased permeability a grave hazard until proven otherwise.

Researchers have noted several other potentially dangerous direct effects of electromagnetic smog on the neurons. In 1980 a group under R. A. Jaffe at Pacific Northwest Laboratories in Richland, Washington, found a general increase in neural excitability, especially at the synapses, in rats exposed to 60-hertz electric fields of only 10 volts per centimeter for one month. That same year A. P. Sanders and co-workers at the Duke University Medical Center in Durham, North Carolina, reported as follows on biochemical tests of rat brains subjected to microwaves at two levels, one half and also slightly more than the U.S. safety standard of 10,000 microwatts: "The results suggest that microwave exposure inhibits electron transport chain function in brain mitochondria and results in decreased energy levels in the brain."

In a series of experiments spanning more than a decade, a group of scientists headed by Ross Adey, first at UCLA and later at the Loma Linda VA Hospital, have studied neuron response to ELF fields and pulses. Proceeding from Hamer's work on reaction time, they first ascertained that an even weaker electric field, roughly the influence of a light bulb 10 feet away, changed the firing rate of brain cells in monkeys and humans *if the field was pulsing at brain wave frequencies.* Then, working with radio waves beamed at chick brains kept alive in culture dishes, they found specific pulse rates that decreased or increased the binding of calcium ions to the nerve cells. The flow of calcium ions in and out of neurons controlled the firing rate of impulses in a complex feedback system. Two "windows" of pulsed radio waves (147 megahertz pulsed at 6 to 10 hertz, and 450 megahertz pulsed at 16 hertz) increased the flow of calcium from the cells, interfering with impulse transmission.

Unfortunately for conceptual simplicity but fortunately for the test animals and the rest of us, the pulsed frequencies that work on isolated brains don't work on whole animals. Adey has publicly expressed his conviction that pulses for changing calcium flow in intact nervous systems do exist, however, and he expects that a calcium efflux would interfere with concentration on complex tasks, disrupt sleep patterns, and change brain function in other ways that can't be predicted yet. This research obviously points toward "confusion beam" weaponry, so effective windows may already have been found, but they haven't been reported in the open literature. Be that as it may, Adey's work remains an important clue to the interaction between EMR and the human CNS at the brain's most sensitive frequencies. Together with the other findings just mentioned, it shows that electropollution can trigger profound and dangerous changes, even if we don't yet know exactly how and when.

Just how dangerous these changes may be was indicated by a study that Maria Reichmanis, Andy Marino, and I did in 1979, collaborating with F. Stephen Perry, a doctor near the town of Wolverhampton in western England. Perry had noticed that people living near overhead high-voltage lines seemed more prone to depression than others in his practice. Since ELF electric fields changed norepinephrine levels in rat brains and since depletion of this neurotransmitter in certain brain areas was a clinical sign of depression, the connection seemed plausible. We knew from earlier work that, although electromagnetic field strength fell off quickly in the immediate vicinity of a power line, the rate of decrease lessened with distance, so that the field was often well above background levels over a mile away. Reasoning that suicide was the one unequivocal and measurable sign of extreme depression, we plotted the addresses of 598 suicides on maps showing the location of power lines in Perry's locality. Then we statistically compared this distribution with a set of addresses chosen at random.

The suicide addresses were, on the average, closer to the high-voltage wires. We found the same association with underground power lines, but we couldn't be sure whether more than the statistically expected number of suicides had occurred in areas where the fields were strongest. Since the total field strength was a combination of elements from many sources, we proceeded to measure the actual levels. This confirmed a link. Magnetic fields averaged 22 percent higher at suicide addresses than at the controls, and areas with the strongest fields contained 40 percent more fatal locations than randomly selected houses.

The Endocrine, Metabolic, and Cardiovascular Systems

Living things interpret electromagnetic energy for information about time and place, so they must have a means to filter out useless signals, although perhaps not those never before encountered. Many studies have found that the bioeffects of artificial energy stabilize after a few weeks, suggesting that animals adapt so as to live normally in a changed environment. Hence there's a large body of work that's often quoted to "prove" that electropollution isn't dangerous. As already noted, this simplistic viewpoint doesn't take into account the additive effects of stress. Moreover, when a stress is too strong or too persistent, compensation fails, and the effects become obvious and sometimes irreversible. When evaluating research on hazards, therefore, we must always ask whether the experiment was continued long enough to be informative. Otherwise, a short-term study showing harm is likely to be truer than a reassuring one of medium length.

The primary effect of electromagnetic energy on the endocrine system appears to be the stress responses already described. The major confirmatory study in humans comes from the Soviet Union, where detailed medical tests of seventy-two technicians exposed daily to 1,000 microwatts or less disclosed ominous changes in white and red blood cell counts and an across-the-board decline in immune response. The workers and a group of controls were studied for three years. No human study approaching this in length or completeness has ever been done in the West.

The only other consistently noted glandular change is in the thyroid. The work of several Soviet groups and one American team in the 1970s has clearly shown that radio and microwave frequencies, at power densities well below the American safety guideline of 10,000 microwatts, stimulate the thyroid gland and thus increase the basal metabolic rate. ELF fields at 50 hertz, on the other hand, have depressed thyroid activity in several experiments on rats. It isn't yet known whether this is a direct effect on the thyroid or whether, like the stress response, it's at least partly caused by alterations in brain function.

One more link in the bioclock-interference-and-stress response has come from 1980 work at the Battelle Pacific Northwest Laboratory in Richland, Washington. Working with rats, researchers there found that a weak 60-hertz electric field (only 3.9 volts per centimeter) canceled the normal nightly rise in production of the pineal gland hormone melatonin, the main hormonal mediator of biocycles.

The cardiovascular system responds to electromagnetic energy in at least two ways. The composition of the blood reflects the stress response and concomitant activation of the immune system, while many frequencies exert a direct effect on the electrical system of the heart.

Soviet scientists have observed a variety of blood changes in animals exposed to microwaves, radio waves, and ELF fields. These include declines in red blood cell count and hemoglobin concentration—and hence oxygen capacity—as well as changes in the relative numbers of various types of white blood cells and the relative amounts of blood proteins, and a possible reduction in the blood's ability to clot.

Most of the discomforting studies of electropollution have been done by the Soviets, and they've been given short shrift by Western scientists. There are many reasons for this attitude. There's a simple prejudice against all things Russian and a feeling that their science, technologically less flashy than ours, is necessarily cruder. Western researchers have hamstrung themselves with the dogma that there simply *can't be* bioeffects from low levels of electromagnetic energy—so why bother looking? Then, too, Russian publication standards are different; procedural details are often omitted, making replication difficult. In addition, there are often troubling contradictions in the data themselves. Results are often inconsistent from animal to animal. If red blood cell count goes up in one, it will go down in another, so the experiment shows no statistical change even though every animal's blood composition is going haywire. In such a situation, the ultramechanistic Americans tend to believe the statistics, while the Soviet biologists concentrate on the animals. Russian scientists have been systematically studying electromagnetic bioeffects since 1933, and we can hardly afford to dismiss their entire body of work simply because it comes from a country we fear.

My associates and I therefore proceeded from one of the most detailed Soviet reports and designed an experiment to measure effects on the blood of mice as our test fields were turned on and off. We concluded that these effects weren't a reaction to the fields themselves but rather a transient compensation that the animals were making in response to *any change* in their electromagnetic environment. By themselves, none of the blood fluctuations were especially hazardous. However, since we all live amid EMFs that are constantly shifting as we turn appliances on and off or travel from place to place, the continual blood instability could be significant.

American attitudes began to alter in 1978–79 when Richard Lovely of the University of Washington took advantage of a detente-inspired exchange of microwave results to visit the Soviet Union for a month and study Eastern methods closely. His research group then painstakingly re-created a major Soviet experiment in which rats had been irradiated seven hours a day for three months with 500 microwatts. The Russian work was confirmed in every detail, including disruption of the blood's sodium-potassium balance, other pathological changes in blood chemistry, damage to the adrenal glands from stress-induced hormonal changes, diminished sense of touch, a decline in explorativeness, and slower learning of conditioned responses. Donald I. McRee, director of the EPA electromagnetic-radiation health research program, termed the results "very interesting" and called for an end to the American establishment's contempt for Soviet work.

Electromagnetic energy has other adverse effects on blood composition and tissue function. Yuri D. Dumansky, one of many Soviet bio-physicists who have done detailed work on microwave hazards, found changes in carbohydrate metabolism, including a rise in human blood-sugar levels, resulting from 100 and 1,000 microwatts. Power-frequency (50-hertz) fields were also linked to altered sugar and protein metabolism in rats, as well as decreased muscular strength in rabbits. Like many other Russian results, these were questioned because of American failure to corroborate them. In this case a research team headed by N. S. Mathewson of the Armed Forces Radiobiology Research Institute in Bethesda, Maryland, reported no such metabolic changes in response to the Sanguine-Seafarer 45-hertz frequency.

However, the Mathewson group made a fundamental mistake. They neglected to account for the 60-hertz background field near the test cages in their lab, even though they'd measured it when setting up the work station. When we reanalyzed their data in light of this omission, the experiment showed exactly the same changes in blood levels of glucose, globulins, lipids, and triglycerides as the Russians had found.

The most frightening data so far on blood composition come from a preliminary study for Project Sanguine. Dietrich Beischer found that *one day* of exposure to a magnetic field such as would be produced by the

ELF antenna caused a *50-percent* rise in triglycerides in nine of ten human subjects. The NAS committee's conclusion that this early result didn't stand up was based on subsequent Navy work, mainly the faulty Mathewson study. Adequate follow-up by a disinterested group has never been funded, even though Beischer's finding agrees completely with Russian studies and the reinterpreted Mathewson data on animals.

This doesn't exhaust the list of microwave metabolic effects reported behind the Iron Curtain. Dumansky found widespread changes in the liver function of rats exposed to low levels of microwaves that were scheduled to approximate the pattern of mealtime exposure from microwave ovens. Others detected vitamin B_2 and B_6 depletion from blood, brain, liver, kidneys, and heart, as well as major shifts in trace-metal metabolism in response to low levels of microwaves. The distribution of copper, manganese, molybdenum, nickel, and iron was affected throughout the bodies of rats. Similar trace-metal changes were recorded after exposure to ELF electric fields for four months, even at moderate field strengths for only half an hour per day. Since B_6 is essential to the utilization of carbohydrates, fat, and protein, and since the trace metals act as catalysts in a wide variety of biochemical reactions, these observations may explain some of the other metabolic changes.

There are indications that some types of electropollution directly decrease the efficiency of the heart. Several research groups in Poland, the Soviet Union, Italy, and the United States have studied pulse, electrocardiogram, blood pressure, and reserve capacity (the heart's ability to handle exertion) in animals. Microwaves and 50-hertz electric fields both produced similar changes that persisted throughout long-term exposure. These included bradycardia (decreased pulse), a huge reduction (40 to 50 percent) in the strength of the electrical impulses governing contraction of the heart muscle, a decline in reserve capacity, and a short-term rise followed by a long-term fall in blood pressure. In general, these decrements occurred in both "domestic" (0.5 volts per centimeter) and "industrial" (50 volts per centimeter or more) electric fields and at microwave power densities of 150 microwatts, well within the amount received by many people from radar beams and microwave ovens.

In humans, confirmatory evidence for these effects comes from several Russian studies of workers in high-voltage power station switchyards. In the first such group examined, forty-one of forty-five had some sign of nervous or cardiovascular disease, including bradycardia, instability of pulse and blood pressure, and tremors. The same health problems were found in four additional studies of nearly seven hundred more workers.

The only comparable American study is much quoted for its failure to find consistent health damage in a mere eleven power-line maintenance workers.

Growth Systems and Immune Response

Given the results presented so far and the dynamics of life's connection to the earth's field, we can now make several predictions about the effects of ELF pollution. The most important aspects of the natural electromagnetic field for the biological timing systems are the lunar circadian rhythm and the micropulsations of 0.1 to 35 hertz. It seems logical that cells will perceive frequencies close to normal more readily than those further removed from the norm. Therefore we can postulate that the ELF band from 35 to 100 hertz would be the most damaging, while higher frequencies might go more or less unnoticed until the energy injected into cells became intense or prolonged enough to be significant. The accumulating evidence supports this idea.

Based on this notion, we can predict two major ELF effects that would encompass many others. We can expect the abnormal signals to disrupt biocycles. Such disruption would trigger the generalized stress response even if the EMR-induced changes in brain neurotransmitters were only an effect and not a cause of the stress reaction. In addition, the wrong timing signals would likely throw off the mitotic cycle time of every cell, interfering with growth processes throughout the body.

Although any number of factors can trigger the adrenocortical stress reaction, the response itself is always the same. It involves the release from the adrenal glands of specific hormones, mainly the corticosteroids, which in turn mobilize the body against invading germs or foreign proteins. Thus the stress response always activates the immune system.

Short exposures to stress aren't necessarily harmful and may even be healthy. In fact, the Soviet work on microwave stress has disclosed a brief period of increased immune-system competence at very low intensities (under 10 microwatts). However, when an organism must face a continual or repeated stress, the response system enters the chronic phase, during which resistance declines below normal and eventually becomes exhausted. Several well-known diseases, such as peptic ulcer and hypertension, result directly from this stage, but the most important result is a decrease in the body's ability to fight infection and cancer.

The trouble is that the immune system is geared to fight tangible invaders—bacteria, viruses, toxins, and misbehaving cells of the body itself—or such consciously detectable stresses as heat, cold, or injury. It includes a system of circulating antibodies by which specialized cells recognize the intruder. The cells controlling this phase, which is called humoral immunity, then select appropriate defenders from an array of other types, each programmed for a certain function, such as digesting bacteria, clearing away cellular debris, or neutralizing poisons. Electromagnetic energy isn't consciously perceived, however. It tricks the immune system into fighting a shadow. Thus we can predict that, just like a fire company answering a false alarm, the body will be less able to fight a real fire.

Experiments bear out this supposition. Impaired immune response has been found at many frequencies. Several groups of Soviet researchers have found a decline in the efficiency of white blood cells in rats and guinea pigs after the animals had been exposed to radio waves and microwaves. Most of these experimenters checked for immune system disruption only up to power densities of about 500 microwatts, one twentieth of the nominal American safety standard. Multiple dangers from higher levels are already considered proven in the Soviet Union.

As predicted, however, the most dramatic reported effect on immune response has been produced by ELF fields. During his systematic study of 200-gauss, 50-hertz magnetic fields, Yuri N. Udintsev found that the concentration of bacteria needed to kill mice in such an environment was only one fifth that needed without the field.

When considering resistance to illness, we must also account for the effect of electromagnetic energy on the disease itself, a factor that has so far been all but ignored. Virtually the only evidence to date is a disturbing piece of work by Yu. N. Achkasova and her colleagues at the Crimean Medical Institute in Simferopol. In 1978 they reported the results of exposing thirteen standard strains of bacteria—including anthrax, typhus, pneumonia, and staphylococcus—to electric and magnetic fields. After accounting for magnetic storms, ionospheric flux, passage of the interplanetary magnetic-field boundaries, and other variables, they found clear evidence that an electric field only slightly stronger than earth's background stimulated growth of all bacteria and increased their resistance to antibiotics. The magnetic fields inhibited the growth of the germs but in many cases still enhanced their resistance to antibiotics. Achkasova concentrated on frequencies between 0.1 and 1 hertz, so the survey was far from complete, but perhaps the most important finding was that *every* field tested had an effect, even after one four-hour exposure. In many cases longer exposure produced *permanent* changes in bacterial metabolism.

The admittedly sketchy evidence to date suggests that our electropollution is presenting us, and perhaps all animals, with a double
challenge: weaker immune systems and stronger diseases. We shouldn't
be surprised, then, at an onslaught of "new" ailments, beginning about
1950 and accelerating toward the future. In several cases, new maladies
have recently been described as coming from pathogens that previously
weren't capable of inducing disease, and this, too, shouldn't surprise us.
Among the newcomers are:

- *Reye's syndrome.* First described in 1963, this condition begins with
  severe vomiting as a child is recovering from the flu or chicken
  pox. It then progresses to lethargy, personality changes, convulsions, coma, and death. The mortality rate, initially very high,
  has now been reduced to about 10 percent, but the incidence has
  increased greatly.

- *Lyme disease.* A virus disease carried by certain insects, it produces
  severe arthritis in humans. It's one of several similar illnesses that
  have appeared only recently.

- *Legionnaire's disease.* This is a pneumonia caused by a common soil
  bacterium that has found a second home in air-conditioning systems. The organism caused us no recognized problems before the
  initial outbreak in Philadelphia in 1976.

- *AIDS.* Autoimmune deficiency syndrome is a condition in which
  the body's immune system fails completely and its owner often
  dies. The patient is unable to resist common, otherwise harmless
  bacteria and viruses, and can no longer suppress the seeds of cancer that reside in all of us. At present, some sort of virus is suspected as the precipitating cause.

- *Herpes genitalis.* This disease isn't new, but its prevalence and severity
  have increased tremendously in one decade. Sexual permissiveness
  generally takes the blame, but a decline in immunocompetence may
  be more important.

Certainly there are additional factors that may be contributing to the
rise of these and other new illnesses. Chemical pollution and the prevalence of junk food are two of the most obvious. However, these diseases,
as well as cancer, birth defects, and the other growth problems described
below, are on the increase throughout the industrialized world. So are
some of the major psychological diseases, such as depression and compulsive use of all types of drugs, from caffeine, nicotine, and alcohol to
prescription tranquilizers and the illegal euphoriants. Although heart-
attack death rates have declined in the last five years (for no known

reason), they're still far higher than before World War II. These diseases exist at more or less the same rates in countries whose chemical toxicity, eating habits, and styles of life are widely divergent. However, the massive use of electromagnetic energy is a common denominator uniting all of the developed nations. In particular, the entire North American continent, Western Europe, and Japan generate such strong 50- and 60-hertz fields that they can be sensed by satellites in space. The populations of these areas are continuously bombarded by these ELF fields.

Disruption of the biocycle timing cues must inevitably make it harder for the body to regulate the mitotic rate of its cells. The major exception to the "no effect" assurances in the NAS Sanguine-Seafarer report was unignorable evidence that 75-hertz fields lengthened the mitotic cycle and hindered cell respiration of the slime mold used in standard tests of cellular growth. The same effects were seen regardless of field strength. Hence we should expect that ELF pollution would foster diseases in which growth processes go awry.

Indeed there has been an alarming increase in such problems. Cancer is hardly a novel illness, but its prevalence is new. In the mid-1960s roughly a quarter of the U.S. population could expect to develop it. By the mid-1970s, that figure had risen to one third, and it's now even higher. The incidence of birth defects has doubled in the past quarter century. There has been a similarly rapid rise in infertility and other reproductive problems.

Rarer defects of cell division may be on the increase as well, expecially among workers exposed by occupation to high levels of electromagnetic energy. Pathologist Hylar Friedman of the Army Medical Center in El Paso reported in 1981 that radar technicians were three to twelve times more likely than the rest of the population to get polycythemia, a rare blood disorder characterized by production of too many red blood cells. Such relationships are hard to confirm statistically, however, in diseases affecting small numbers of people. We need direct experimental evidence and large-scale studies on the widespread disorders. Both are now available.

Back in 1971, two more Soviet researchers, S. G. Mamontov and L. N. Ivanova, reported that industrial-strength 50-hertz electric fields tripled the mitotic rate of liver and cornea cells in mice. Soon afterward, Bassett and Pilla published empirical evidence that pulsed EMFs accelerated the healing of bone fractures. For the most part, however, concrete evidence that time-varying fields could affect cell division was slow in coming.

That situation has changed in the last few years. Several experimenters, notably Stephen Smith, have now proven that the Electrobiology bone-healing device, using 15 pulse-bursts per second, speeds up the division rate of cells that are already proliferating rapidly. Among normal cells, this includes skin, gut, and liver cells. In 1983, A. R. Liboff, a biophysicist at Oakland University in Rochester, Michigan, reported on the effects of a more inclusive set of parameters. Magnetic fields of 0.2 to 4 gauss, vibrating at 10 to 4,000 hertz, all enhanced the replication of DNA during the S (synthesis) phase of mitosis.

As predicted, the interaction appears to be greatest between 35 and 100 hertz. José M. R. Delgado—the flamboyant advocate of a "psychocivilized" society through mind control, who has publicized direct electrical stimulation of the brain by such displays as stopping a charging bull in its tracks with a radio impulse transmitted to an implanted electrode—recently reported results of a genetic study of magnetic fields at three frequencies. Delgado placed chick embryos in minuscule magnetic fields pulsed at 10, 100, and 1,000 hertz. He used fields of only 0.001 gauss, or roughly the strength of the earth field's micropulsations. Chicks exposed to the 10-hertz fields were normal, but those dosed at 100 hertz developed severe defects of the central nervous system. The highest frequency also yielded abnormalities, but they were much less severe. Higher intensities are common in homes, in offices, and near power lines. The Navy has found stronger fields near its 76-H_3 ELF antenna and reradiated at that frequency from a power line a mile away.

It's important to bear in mind that, in stimulating DNA synthesis, an electromagnetic field doesn't distinguish between desirable and undesirable growth. It affects all cells in the same way, but cell systems that are already rapidly dividing are speeded up the most. As we've seen in earlier chapters, these susceptible processes include healing, embryonic growth, and cancer.* In fact, a researcher working on the New York State Department of Health's power line project, Wendell Winters of the University of Texas Health Sciences Center in San Antonio, recently reported some of the first laboratory evidence that power frequencies can accelerate malignant growth. Winters exposed human cancer cells to 60-hertz electromagnetic fields for just twenty-four hours, and found a sixfold increase in their growth rate seven to ten days later.

---

*Only the magnetic component appears to accelerate healing in any way. Power-frequency electric fields severely retard fracture healing in rats, as Andy Marino, Jim Cullen, Maria Reichmanis, and I proved with a series of experiments in 1979. This work was confirmed the following year by R. D. Phillips in a study done for a Department of Energy review of transmission line bioeffects.

Moreover, the perturbation of normal cell-cycle time is enhanced if nuclear magnetic resonance (NMR) is induced in the atoms of the DNA molecules. In simplified terms, nuclear magnetic resonance is present when the magnetic fields around atomic nuclei are induced to vibrate in unison. The phenomenon requires two external magnetic fields, one steady and one pulsating. For every chemical element, the oscillating field at a specific frequency will induce resonance within the steady-state field at a certain strength.

In 1983 a research team under A. H. Jafary-Asl showed that the earth's magnetic background could serve as the steady field, while the harmonics of power line frequencies could produce a time-varying field that would induce nuclear magnetic resonance in at least two common atoms of living tissue—potassium and chlorine. Other elements might also be susceptible to the effect. Bacteria and yeast cells exposed to these NMR conditions doubled their rate of DNA synthesis and proliferation, but daughter cells were half size. Liboff, analyzing contradictory studies, found that the contradictions disappeared when he calculated resonance conditions for the earth's field where each test was done. Previous work must now be reinterpreted as one vast experiment in adding new frequencies to the varying background.

Almost all experimenters to date have tested the response of organisms to a single specific frequency and intensity. This approach was needed in the beginning to provide a basic level of knowledge, but it's far removed from everyday life, in which we're all exposed to many frequencies simultaneously. A synergism between electromagnetic energy and radioactivity has already been suggested by the fact that cancer rates among nuclear power plant workers are higher than was predicted solely by the higher levels of ionizing radiation in their environment. Nuclear power plants abound in multifrequency radio waves and other electromagnetic radiation. In addition to inducing NMR in the building blocks of living cells, multiple frequencies may likewise interact synergistically to yield biohazards greater than the sum of their individual dangers.

Animal experiments on the risk of cancer and birth defects from electromagnetic energy are scarce, even in the USSR. The little work that has been done was mostly on microwaves. The only well-known American laboratory study of birth defect dangers used pulsed radio waves and found numerous mutations in fruit fly offspring. In 1976 a Russian group dosed rats with 50 and 500 microwatts for one to ten days. When they then studied somatic (nongenital) cells from the animals, they found chromosome defects in astounding numbers. At the higher power density there were five times as many as in the controls, and even at the lower intensity the number continued to increase (to 150 percent of the normal value) for two weeks after the beams were turned off.

A 1979 study directed by Przemyslaw Czerski of the National Research Institute of Mother and Child, in Warsaw, documented increased numbers of damaged chromosomes in the sperm of mice exposed one hour a day for two weeks to microwave intensities ranging from 100 microwatts up to the American safety standard of 10,000 microwatts. An even more discomfiting set of data came from a mid-1970s Russian experiment in which female mice were subjected to small power densities, 10 to 50 microwatts. Throughout this range there was a decrease in the number and size of litters and an increase in developmental problems among the newborn animals. The rate of stillbirths jumped from 1.1 percent at the lowest intensity to 7 percent at the highest.

Alas, human beings are the main experimental animals in this line of research. Those who contend microwaves pose no danger often quote a survey of twenty thousand Korean War veterans completed in 1980 by C. D. Robinette and others for the NAS—National Research Council's Medical Follow-up Agency. Comparing VA medical records of radar technicians and others heavily exposed to microwaves with the records of controls, this group found no increase in the death rate. This finding can't be relied on, however. Most of the controls were radar *operators*, who are exposed to some radiation from radar beams as well as from their consoles. Thus the presumption that they absorbed negligible amounts of EMR just doesn't hold water. In the last few years more reliable epidemiological studies have appeared, showing increased rates of cancer and birth defects among people exposed to higher-than-average levels of electromagnetic energy.

Since microwave broadcasts for television and telephone relays must be in a line of sight to the receivers, there are only a few suitable high locations for the transmitters near each city. Of necessity there's an above-normal concentration of ELF fields and microwave spilloff in that area, possibly leading to a destructive synergism as outlined above. Moreover, since TV is aimed at an audience and phone relay beams at the next station, corridors are set up within which people get more than their share of microwaves.

Sentinel Heights, seven miles from downtown Syracuse, is one such transmitter hill. Slightly more than a thousand people live there. From 1974 to 1977 I learned of seven cases of cancer in that area. They were divided into two clusters, in two microwave corridors separated by a shadow zone. This is 55 percent more than the 4.5 cases statistically expected for this population, and there may have been more cases I didn't know about. Obviously, in such a small and unscientific sample the results could have been due to chance, but the ominous implications demanded some more extensive surveys.

The first one came in 1979, when Nancy Wertheimer and Ed Leeper of the University of Colorado Medical Center in Denver published a study of childhood cancer and power lines. The researchers studied 344 deaths from childhood cancer between 1950 and 1973. The address of each of the victims was paired with the address of the next baby born in the area, to provide a matched series of controls. If the family had moved before the death, both birth and death addresses were used in the experimental group. The wiring of each house and its distance from the nearest transformers were studied. It proved possible to divide the houses into two groups: those with high-current wiring configurations producing strong magnetic fields, and those wired in a low-current arrangement producing much weaker magnetic fields. After certain other variables—such as economic class, family risk patterns, traffic, and urbanization differences—were factored out, the childhood death rate from leukemia, lymph node cancer, and nervous system tumors in the high-current homes was more than double the rate in low-current homes.

Three years later S. Milham, director of occupational health and safety for the state of Washington, found that adults who worked in strong electromagnetic fields also had a leukemia incidence significantly higher than the norm. The link appeared in statistics for generating-station operators, high-voltage-line maintenance workers, aluminum smelters, and several other categories of laborers.

Besides the investigation itself, another thing was noteworthy about Milham's paper: the reaction of the scientific establishment. Another paper quickly appeared in the same periodical, the *New England Journal of Medicine,* citing many other studies to prove Milham wrong. However, all of them involved controlled exposure to microwaves alone, while the jobs studied by Milham were in the real world, where microwaves and power-frequency fields mix. The editors declined to publish my letter pointing out this obvious flaw in the critique, but still it was momentous that such a prestigious publication ran Milham's paper at all.

Soon confirmatory reports appeared. Wertheimer's and Leeper's findings were duplicated in Stockholm by a group who correlated childhood leukemia with actual measurements of magnetic fields. The strongest statistical link was found with 200-kilovolt power lines running within 200 yards of the stricken child's home. Milham's work was vindicated by surveys in Los Angeles and Great Britain. Wertheimer herself extended her observations to adults and found the same highly significant connection between high-current wiring and various cancers, especially leukemia.

Radar beams (composed of pulsed microwaves) have the highest power densities of any EMR source. In the laboratory, both radio frequency and microwave radiation have been shown to change the gateway-barrier function of cell membranes, upset hormone balances, and induce chromosome defects, all of which are factors in malignant growth. However, there have been few attempts to directly assess radar's potential role in human cancer.

John R. Lester and Dennis F. Moore of the University of Kansas School of Medicine in Wichita have recently done so. Wichita was an ideal location for such an inquiry. It had two airports with radar towers, but few other major sources of electropollution. Its chemical environment was also quite clean as cities go. Lester and Moore plotted the cancer incidence for the whole city and found it was highest where the residents were exposed to both radar beams. It was lower where only one beam penetrated, but lowest where the population was fully shielded behind hills. The results held up when other factors, such as age, poverty, sex, and race, were statistically balanced as far as possible. The authors noted one apartment house whose cancer death rate was twice that of the area's nursing homes; its upper floors were in direct line with both radar beams.

Heart attack rates in North Karelia and Kuopio, Finland, became the highest (and most swiftly increasing) in the world within a few years after the Soviets installed a gigantic over-the-horizon radar complex that bounced microwaves off the surface of Lake Ladoga and through these parts of southeastern Finland. These are rural districts whose way of life is built on outdoor labor rather than the sedentary indoor stresses generally associated with heart disease. Noting that cancer rates had also risen precipitously in the region, Lester and Moore went on to investigate statistics for American counties having Air Force bases. These counties had a significantly higher percentage of cancer deaths than other counties, even though radar towers from commercial airports inevitably must have smoothed out the data and made the difference less striking.

The study of human genetic defects from electromagnetic energy is still in a primitive stage. In the case of microwaves, this situation is largely due to obstruction by military and government agencies. Even in World War II, rumors of radar-induced sterility were so rampant that sailors often gave themselves "treatments" before shore leave. The first scientific evidence of reproductive effects didn't come until 1959, when John H. Heller and his co-workers at the New England Institute for Medical Research in Ridgefield, Connecticut, found major chromosome abnormalities in garlic shoots irradiated with low levels of microwaves. They soon found the same changes in mammalian cells, as well as the fruit fly mutations mentioned above. Their work in this direction ended about 1970 due to lack of funds.

In 1964 a group of researchers studying Down's syndrome at the Johns Hopkins School of Medicine, after linking the malady to excess X rays given to pregnant women, found an unexpected further correlation with fathers working near radar. It was a full decade before any money was allocated to follow up this finding, and, while the link between parental radar exposure and Down's syndrome wasn't substantiated, higher-than-normal numbers of chromosome defects *were* found in the blood cells of radarmen.

By this time an Alabama professor of public health had found an apparent surge in birth defects among children of radar-exposed Army helicopter pilots. In 1971 Dr. Peter Peacock noted that there had been seventeen children born with clubfoot within a sixteen-month period at the Fort Rucker, Alabama, base hospital. Statistically, there should have been no more than four.

Working through two federal agencies and two private research foundations, Peacock and others tried for five years to follow up this disturbing news, only to be thwarted by some clever tactical moves by the Army. Refusing to release work records, medical files, and radar inspection records on grounds of "privacy" and "national security," officials of the Army Medical Research and Development Command managed to prevent all but two reassessments of Peacock's original data for several years. They stalled separate research proposals sponsored by the Environmental Protection Agency and the Food and Drug Administration's Bureau of Radiological Health without ever letting on to one agency that they were dealing with the other. As the coup de grâce, the Army agreed to supply the FDA group with a survey of radar transmitters in the Fort Rucker area. The officers fobbed off on the unwitting civilians a deceitfully sketchy map showing only one major radar installation at the base, whereas an official Army report made at the time of the observed birth defects showed nineteen such emitters. Throughout the Vietnam War thousands of helicopter trainees had each spent months flying through the resultant microwave haze. Much of their training consisted of homing right down the beams to within a few dozen yards of the source in TH-13 Bell copters whose Plexiglas bubbles left them naked to microwaves.

The Fort Rucker affair and many other instances of military-governmental sabotage of health effects research on microwaves have been impeccably documented in *New Yorker* reporter Paul Brodeur's 1977 book, *The Zapping of America.* In the early 1970s, for example, follow-up to a preliminary finding of excess Down's syndrome among children of Seattle airline pilots was first supported by the local chapter of the Air Line Pilots Association, then opposed due to pressure from the national level.

The stonewalling continues. Grants for serious consideration of electropollution's dangers have been cut to a trickle in the United States, but some findings continue to emerge, especially from other countries.

A 1976 survey of Hydro-Quebec's generating-station electricians showed a drastic change in the gender ratio of children born after one of the parents began work in the high-EMF environment. Before, boys and girls had been born in equal numbers; afterward, there were six times as many males as females. A 1979 study of Swedish high-voltage substation workers showed lower birth rates and an 8-percent incidence of genetic defects in offspring, as compared with 3 percent among children of a control group. The finding was confirmed in 1983. Since most of the exposed electrical workers were men, the damage apparently was done during sperm formation. Most recently, in May 1984, Nancy Wertheimer presented evidence of a statistical correlation between use of electric blankets, which emit powerful EMFs, and the occurrence of birth defects.

Among the most serious recent data are those concerning video display terminals (VDTs). There have been alarming numbers of miscarriages, stillbirths, and birth defects among pregnant women working in newly computerized offices. In one year at the Dallas office of Sears, Roebuck and Company, for example, only four of twelve pregnancies ended normally. Among twelve pregnant VDT workers at the Defense Logistics Agency in Marietta, Georgia, there were seven miscarriages and three cases of congenital defects. Four VDT operators in the *Toronto Star*'s classified-ad department gave birth to deformed children, while three co-workers who didn't work with VDTs had normal babies. These anomalies must be compared with the normal 15-percent incidence of spontaneous abortion and the 3-percent rate of serious birth defects among the population at large. Writing in *Microwave News,* an independent newsletter covering nonionizing radiation, in 1982 editors Louis Slesin and Martha Zybko reported on eight such clusters, and workers' groups have documented several others, but still there has been no attempt at a large-scale statistical study to check the oft repeated claim that these are just coincidences.

Two studies are widely quoted as disproving harmful effects from the machines. In 1977, when two *New York Times* copy editors developed radiation-induced cataracts after less than a year at their new screens, the National Institute of Occupational Safety and Health (NIOSH) tested a few machines and, finding that X-ray emissions were within the half-millirem-per-hour standard for work exposure, concluded there was no link to the health problems. Unfortunately, the agency didn't adequately measure nonionizing radiation, gave contradictory data as to the sensitivity of its own instruments, and failed to test *malfunctioning* monitors, which are known to emit larger amounts of X rays. Nor is there any assurance that the X-ray exposure standard is adequate, since it was formulated for a much smaller group of workers (mainly nuclear technicians and uranium miners), whose health is continuously monitored in a way that that of VDT operators is not. Furthermore, the NIOSH investigators noted an enormous microwave reading of 1,000 microwatts in one of the *Times* offices, without even bothering to find out where it was coming from!

Press releases claimed a mid-1983 National Academy of Sciences review would allay the fears once and for all, proving VDTs to be risk free. However, a reading of the text showed a different picture. While the authors played down reports linking birth defects and eye problems to VDT radiation, they admittedly failed to find *any* research adequate to answer the health questions one way or the other.

According to the sketchy data available, all VDTs (which of course include video games and televisions as well as computer monitors) emit varying amounts of radiation over a broad spectrum. The transformers release VLF and ELF waves, while microwaves, X rays, and ultraviolet emanate from the screen. Poorly adjusted or malfunctioning terminals can emit enormous amounts; two machines tested in the offices of Long Island's *Newsday,* for example, were producing 15,000 microwatts of radio energy. There's no information whatever on the synergisms that may operate amid this varied radiation over long periods of time, but I suspect that the birth defects are primarily due to the ELF component.

Meanwhile, the only American "research" on the problem continues to be the daily lives of our 10 million or more console operators. Despite the reassurances, at least a third to a half of the workers continue to suffer headaches, nausea, neck and back pain, and vision impairment. In fact, a 1983 survey of eleven hundred UPI employees conducted by

Arthur Frank, then at New York's Mount Sinai School of Medicine, suggested that VDT users lose so much time due to eye problems and neck pain that the effects may become a major drain on the economy by the end of the decade.

Some of the complaints undoubtedly arise from postural strains and lighting defects in the notoriously ill-designed work areas where many VDTs are used. They could be prevented by more frequent breaks and some sympathetic attention to human engineering. The birth defects and cataracts probably won't disappear so easily, however. Certainly pregnant women should be allowed temporary reassignment without loss of pay, a right already accepted in much of Western Europe and recently put into law in Ontario. That won't protect sperm cells and unfertilized eggs, however. Regular maintenance and a lead-impregnated glass or acrylic screen (such as is used in nuclear power plant windows) can virtually eliminate ionizing radiation, but screen-generated microwaves require a transparent shield that still conducts electrical energy—a product that doesn't yet exist. Some frequencies of EMR are easy to block simply by using metal cabinets instead of the cheaper plastic ones, but VLF and ELF waves require grounded shielding. All these preventive measures are expenses that most manufacturers and managers have been loath to accept; until they do, workers will be paying the entire price.

The dangers of electropollution are real and well documented. It changes, often pathologically, every biological system. What we don't know is *exactly* how serious these changes are, for how many people. The longer we, as a society, put off a search for that knowledge, the greater the damage is likely to be and the harder it will be to correct. Meanwhile, one of the few honest statements to emerge from the Nixon administration, a warning issued by the President's Office of Telecommunications Policy in 1971, continues to bleed through the whitewash: "The population at risk is not really known; it may be special groups; it may well be the entire population. . . . The consequences of undervaluing or misjudging the biological effects of long-term, low-level exposure could become a critical problem for the public health, especially if genetic effects are involved."

Conflicting Standards

The establishment attitude toward EMR's health effects derives largely from the work of Herman Schwan. An engineer who had been a professor at the Kaiser Wilhelm Institute of Biophysics in Germany during most of the Nazi era, Schwan was admitted to the United States in 1947, soon accepted a post at the University of Pennsylvania, and since then has done most of his research for the Department of Defense.

Like infrared radiation, radio waves and microwaves produce heat when they're absorbed in sufficient quantity. Although not a biologist, Schwan assumed this heating was the only effect EMR would have on living tissue. In this respect he considered living things no different from the hot dogs that World War II radarmen used to roast in their microwave beams, so cooking was the only harm he foresaw. Schwan then estimated danger levels based on how much energy was needed to measurably heat metal balls and beakers of salt water, which he used to represent the size and presumed electrical characteristics of various animals.

Appreciable heating occurred in these models only at levels of 100,000 microwatts or above, so, incorporating a safety factor of ten, Schwan in 1953 proposed an exposure limit of 10,000 microwatts for humans. By showing soon afterward that it took more than this intensity to cause burns in real animals, Sol Michaelson seemed to have confirmed the safety of "nonthermal" dosages. No one tested for subtler effects, and the 10,000-microwatt level was uncritically accepted on an informal basis by industry and the military. In 1965 the Army and Air Force formally adopted the Schwan limit, and a year later the industry-sponsored American National Standards Institute recommended it as a guideline for worker safety.

There were persuasive economic reasons why the 10,000-microwatt standard was and still is defended at all costs. Lowering it would have curtailed the expansion of military EMR use and cut into the profits of the corporations that supplied the hardware. A reduced standard now would constitute an admission that the old one was unsafe, leading to liability for damage claims from ex-GIs and industrial workers. One of the strongest monetary reasons was given in a 1975 classified summary of the DOD's Tri-Service Electromagnetic Radiation Bioeffects Research Plan: "These [lower] standards will significantly restrict the military use of EMR in a peacetime environment and require the procurement of substantial real estate around ground-based EMR emitters to provide buffer zones." The needed real estate was estimated to be 498,000 acres. The price of this much land would surely run well into the billions of dollars.

Even before it was adopted, there were indications that the standard might be inadequate. During the obligatory fight for compensation in the face of callous official denials of responsibility, an interesting discovery was made by Thomas Montgomery, a former civilian technician working for the Army Signal Corps, who is now blind, deaf, and crippled because of a massive accidental exposure to a radar beam in 1949. In one of the files opened by his suit, Montgomery found a document proving that in the late 1940s the Institute of Radio Engineers had formulated more conservative safeguards that included methods for preventing accidents like the one that had incapacitated him. (He'd been repairing a transmitter when a co-worker, not knowing he was standing in front of the wave guide, turned it on. Since the microwaves were imperceptible, Montgomery didn't know he was being irradiated until it was too late.) Leaders of the military-industrial electronics community chose not to promulgate these proposed regulations.

There were other hints that all was not well. In 1952 Dr. Frederic G. Hirsch of the Sandia Corporation, a maker of missile guidance systems, reported the first known case of cataracts in a microwave technician. The following year Bell Laboratories, alarmed by reports of sterility and baldness among its own workers as well as military radar personnel, suggested a safety level of 100 microwatts, a hundred times less than Schwan's. Even Schwan has consistently maintained that his dosage limit probably isn't safe for more than an hour.

In 1954 a study of 226 microwave-exposed employees at Lockheed's Burbank factory was reported by company doctor Charles Barron. He said there were no adverse effects, despite "paradoxical and difficult to interpret" changes in white blood cell counts, which he later ascribed to laboratory error, as well as a high incidence of eye pathology, which he determined was "unrelated" to radar.

However, the safety standard had already become a Procrustean bed against which all research proposals and findings were measured. Grants weren't given to look for low-level hazards, and scientists who did find such effects were cut down to size. Funds for their work were quickly shut off and vicious personal attacks undermined their reputations. Later, when undeniable biological changes began to be noted from power densities between 1,000 and 10,000 microwatts, the idea of "differential heating"—hot spots in especially absorptive or poorly cooled tissues—was advanced, as though this convenient explanation obviated all danger. Soviet research could easily be discounted because of its "crudity," but when nonthermal dangers were documented in America, military and industrial spokespeople simply refused to acknowledge them, lying to Congress and the public. Many scientists, who naturally wanted to continue working, went along with the charade.

The pattern was exactly the same as the one regarding fallout dangers from nuclear weapons tests. Throughout the 1950s there was "no cause for alarm," but twenty years later the Wyoming sheep ranchers' suit for compensation for fallout-damaged herds unearthed documents proving the responsible officials had known better at the time. Even the symbol of American military machismo may have fallen victim to the policy. John Wayne, as well as Susan Hayward and other cast members, died of cancer about two decades after making a movie called *The Conqueror,* which was filmed in the Nevada desert while an unexpected wind shift sifted radioactive dust down on them from a nearby test.

Today the EMR deceit still proceeds. On August 2, 1983, Sol Michaelson was quoted as saying some bioeffects had been observed in animals and a few claimed in humans from intensities under 10,000 microwatts, but "none of these effects, even if substantiated, could be considered hazardous or relevant to man"—even though three years before he'd co-authored a paper that reviewed previous evidence and added some of his own in support of the generalized stress response from microwaves.

The evidence had begun to come in over twenty years previously. John Heller's 1959 finding of chromosome changes in irradiated garlic sprouts and the 1964 Johns Hopkins correlation of Down's syndrome with parental exposure to radar were mentioned in the previous section. In 1961 a study was conducted on a strain of mice bred to be especially susceptible to leukemia and used to evaluate risk factors for that disease. Two hundred mice, all males, were dosed with 100,000 microwatts at radar-pulse frequencies for one year. An unusually high proportion of the animals—35 percent—developed leukemia during that time, and 40 percent suffered degeneration of the testicles. Admittedly, this was a very high power density, but the mice were exposed for only four minutes a day. The most disturbing part, however, is that the sponsor, the Air Force, cut off all funds for follow-up work, and to this day no American research has adequately addressed this potential danger.

In 1959 Milton Zaret, an ophthalmologist from Scarsdale, New York, began a study for the Air Force to see if there was any special risk to the eyes of radar maintenance men. He at first found none, because he'd geared the examinations to check the *lens* of the eye. A few years later, when several private companies referred to him microwave workers who'd developed cataracts, Zaret saw he'd made a mistake. Because the microwaves penetrated deeply into tissue, cataracts from them had developed *behind* the lens, in the posterior capsule, or rear part of the elastic membrane surrounding the lens. At this point the Air Force again suddenly lost interest, but Zaret has doggedly pursued the matter in his own practice. Although military and industry people still deny that nonthermal microwave cataracts exist, Zaret's work has proven beyond good-faith dispute that low doses exert a cumulative effect that eventually stiffens and clouds the posterior capsule. Surveys by others here and abroad have confirmed his conclusion. In fact, these behind-the-lens cataracts constitute a "marker disease" for sustained microwave exposure. Zaret has personally diagnosed over fifty clear-cut cases, many in airline pilots and air traffic controllers.

In 1971 Zaret was involved in a projected five-year primate study for the Navy when he made the mistake of telling his overseers the disturbing news that one of the monkeys had died after a few hours' exposure to just twice the U.S. safety level. Within days Captain Paul Tyler "happened to be in the area" of the Hawaii research facility and wanted Zaret to show him around. He left in about twenty minutes, having seen little of the equipment, and in a few days more the entire project had been canceled.

The double-dealing has only made Zaret dig in his heels. Through the years he has been one of the few doctors willing to take on the government by testifying on behalf of plaintiffs filing claims for microwave health damage. At such proceedings one always runs into the same cast of characters speaking more or less the same lines. At one trial the ubiquitous Michaelson attacked Zaret's professional abilities, only to have later testimony reveal the embarrassing fact that Michaelson's mother owed the vision in one eye to the ophthalmologist's surgical skill.

As we've already seen, there are many indications that EMFs and EMR weaker than the Schwan guideline have serious effects on growth. The 10,000-microwatt level was directly tested in 1978 by a group at the Stanford Research Institute. Pregnant squirrel monkeys and their offspring were irradiated. Of nine babies zapped *in utero* and/or after birth, five died within six months, compared with none in the control group.

The U.S. safety standard would be grossly inadequate even if it was law. In actuality, although some businesses and military agencies have adhered to it, it has never posed any threat to those that have not. A federal court case decided in 1975 and upheld in 1977 defined it as an advisory or "should" guideline, which couldn't be enforced. Now, as per instructions dated March 17, 1982, Occupational Safety and Health Administration (OSHA) inspectors can no longer issue even meaningless citations to companies for exposing workers to more than the limit.

The only actual regulation of public electromagnetic energy dosage pertains to microwave ovens. In 1970 the FDA's Bureau of Radiological Health stipulated that no oven should leak more than 1,000 microwatts at a distance of 2 inches when new, nor more than 5,000 after sale. Even at that time, research suggested this was an unsafe level, a fact recognized by Consumers Union in 1973 when it recommended against the purchase of any brand. Leakage surveys have shown that all types put out an average of 120 microwatts near the door, while many emit much higher amounts. A worn seal or piece of paper towel stuck in the door can increase the user's exposure to well over the 5,000-microwatt level, according to Consumers Union tests.

What amounts of electromagnetic energy do workers and the general public actually absorb? The levels vary greatly. Some antenna repairers receive up to 100,000 microwatts for minutes or hours during a job. Many factory workers are in the same bracket. From 1974 to 1978 NIOSH surveyed eighty-two industrial plastic molders and sealers. Over 60 percent exposed the operator to more than the Schwan limit, some to over 260,000 microwatts. Because of low wages for such work, nearly *all* sealer operators are women of childbearing age. NIOSH has estimated that some 21 million workers are exposed to some level of radio-frequency waves or microwaves as a direct result of their jobs. No metal shielding is provided for most workers in this country, although it's sold to other nations having better safety rules.

At this time there's no way to estimate how much EMR people are getting away from their jobs, because the few readings that have been taken have measured only single sources and single frequencies. No one has yet surveyed our cities and countryside throughout the whole spectrum from ELF to microwaves. All we know is that most people's daily exposure is high. Even the Environmental Protection Agency has estimated that if the Soviet off-the-job safety limit of 1 microwatt in the radio and microwave bands was adopted here, over 90 percent of our FM stations would have to be shut down.

Table 1. Power Density at Various Distances from a 50,000 Watt AM Radio Station

Distance (feet)	Power Density ($\mu W/cm^2$)	Distance (feet)	Power Density ($\mu W/cm^2$)
15	838	482	23
29	284	663	12
69	196	1571	2
152	43	3280	1
308	33	5760	0.3

Note: Data from R. Tell et al., "Electric and Magnetic Field Intensities and Associated Body Currents in Man in Close Proximity to a 50 kw AM Standard Broadcast Station," presented at Bioelectromagnetics Symposium, Seattle, 1979.

Table 2. EMF in Typical Tall Buildings

City	Location	Power Density ($\mu W/cm^2$)
New York	102nd Floor, Empire State Building	32.5
Miami	38th Floor, One Biscayne Tower	98.6
Chicago	50th Floor, Sears Building	65.9
Houston	47th Floor, 1100 Milam Building	67.4
San Diego	Roof, Home Tower	180.3

Note: Data from R. Tell and N. H. Hankin, *Measurements of Radio Frequency Field Intensity in Buildings with Close Proximity to Broadcast Systems,* ORP/EAD 78-3, U.S. Environmental Protection Agency, Las Vegas, 1978.

Table 3. Power-Frequency Electric Fields of Household Appliances Measured at a Distance of One Foot

Appliance	Electric Field (v/m)
Electric blanket	250
Broiler	130
Phonograph	90
Refrigerator	60
Food mixer	50
Hairdryer	40
Color TV	30
Vacuum cleaner	16
Electric range	4
Light bulb	2

Note: Data in tables 3 and 4 from *Fact Sheet for the Sanguine System: Final Environmental Impact Statement,* U.S. Navy Electronic Systems Command, 1972.

Table 4. Power-Frequency Magnetic Fields of Household Appliances

Range	Appliance
10–25 gauss	Soldering gun Hairdryer
5–10 gauss	Can opener Electric shaver Kitchen range
1–5 gauss	Food mixer TV
0.1–1.0 gauss	Clothes dryer Vacuum cleaner Heating pad
0.01–0.1 gauss	Lamp Electric iron Dishwasher
0.001–0.1 gauss	Refrigerator

BOSTON 29,300
NEW YORK 49,100
WASHINGTON 70,500
CHICAGO 19,000
LOS ANGELES 7,000
ATLANTA 9,800
MIAMI 29,900

SOURCE: U.S. ENVIRONMENTAL PROTECTION AGENCY, 1978

NUMBER OF PEOPLE IN CERTAIN CITIES EXPOSED TO RADIO AND TV SIGNALS ABOVE USSR SAFETY LEVEL

Most city dwellers continuously get more than a tenth of a microwatt from television microwaves alone. This may be especially significant, because of the human body's resonant frequency. This is the wavelength to which the body responds "as an antenna." Next to the ELF range, it's perhaps the region of the spectrum in which the strongest bioeffects may be expected. The peak human resonant frequency lies right in the middle of the VHF television band.

Many people live in zones of higher-than-average risk. Levels rise steeply from 1 microwatt within half a mile of most radio stations. Near "antenna farms" like the forest of transmitters on Mount Wilson on the outskirts of Los Angeles, the densities can reach well into the thousands. They regularly go up to about 7 microwatts near microwave relay towers, which are often placed in the center of towns. Exposure to 100 microwatts is not uncommon within half a mile of military or airport radar towers. Office workers in tall buildings are often in direct line with microwave beams, whose intensities may reach 30 to 180 microwatts, as measured in a recent EPA survey. CB radios and walkie-talkies bombard users, especially their heads and chests, with thousands of microwatts. These figures, of course, represent only single sources, not the total exposure.

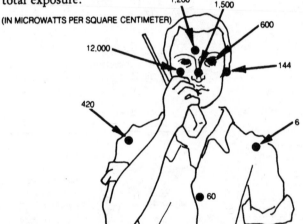

MICROWAVE EXPOSURE FROM WALKIE-TALKIE

Although most people's absorption doesn't approach the Schwan guideline, it should be clear by now that the lower levels are little cause for comfort. Everywhere in Western nations, except in the most remote forests or deserts, the ambient energy from ELF power systems is several thousand times above the earth's background field strength, providing abundant interference with the biocycle timing cues. Moreover, the accumulated research has clearly shown that small doses often have the same effects as larger ones. Ross Adey, who has intensively studied the "window effect," in which a certain result is produced at some frequencies and power levels but not at others interspersed between the effective ones, believes future research will reveal such windows at much lower levels, even at fractions of a microwatt. Indeed there has already been one report of brain wave changes suggesting resonance of neural electrical currents with radio waves and microwaves down to a billionth of a microwatt.

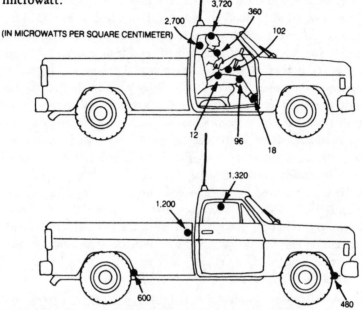

MICROWAVE EXPOSURE FROM A TYPICAL CB RADIO

There's a chance of somewhat stricter rules in a few years. In 1982 the American National Standards Institute recommended that the radio-wave safety level be lowered to 1,000 microwatts and the microwave level to 5,000. This was the first semiofficial admission that nonthermal effects do exist. Now several federal agencies have begun discussing a formal regulation. The most likely source is the Environmental Protection Agency, but at last report the rumored EPA proposal of a 100-microwatt limit for the general public, which had been anticipated in late 1984, was abruptly and indefinitely postponed due to dissension within and pressure from outside the agency.

A health-protecting federal standard with the force of law would have a major impact on both industry and government. Industry would experience a decline in revenue and an increase in costs. Government, especially the military, would be inconvenienced in a multitude of activities. Both would be subject to lawsuits for exposures and damages prior to establishment of the standard. In addition, we must understand that *no* amount of artificial EMR, no matter how small, has been proven safe for continuous exposure. Bioeffects have been found at the lowest measurable doses. However, we must also understand that the greatest danger lies in uncontrolled exposure to large amounts of EMR at many overlapping frequencies, and therefore a stringent standard with a definite time-table for phasing it in is the only way to protect the public health.

Moreover, such action must come from Washington. New Jersey and Connecticut have recently adopted the ANSI standard, while in 1983 Massachusetts enacted a much stricter one of 200 microwatts, which large areas of New York City already exceed considerably. Some communities, recognizing that even a 100-microwatt level is too high, are beginning to set their own, lower, ones. Without *realistic* federal regulation we will end up with a totally unworkable patchwork. Suppose, for example, that the Air Force, from a base *outside* a town, operates a radar dome that produces illegal EMR levels *inside* it. Without federal direction, that will become one more confused legal issue to be hammered out for years in already overburdened courts.

All of the industrialized West is locked into a false position on electropollution's risks. It's these countries that have made the maximum use of electromagnetism for power, communications, and entertainment. The Soviet Union and China, partly due to underdevelopment and wartime destruction, and partly by choice, have severely limited its use and the exposure of their civilians.

Soviet scientists have consistently assumed that *any* radiation that doesn't occur in nature will have some effect on life. We've consistently made the opposite assumption. Throughout our recent history American regulators have followed a "dead body policy." They have extended no protection until there was proof of harm sufficient to overcome all deception. There's no longer any question that, as far as electromagnetic energy is concerned, we've been wrong and the Soviets have been right.

In the 1950s, Russian doctors conducted extensive clinical exams of thousands of workers who had been exposed to microwaves during the development of radar. Having disclosed serious health problems, these studies weren't swept under the rug. Instead, the USSR set limits of 10 microwatts for workers and military personnel, and 1 microwatt for others. Both levels are strictly enforced. When this first became known in the West in the early 1960s, instead of checking their assumptions many American scientists and administrators chose to believe this was Russian propaganda aimed at embarrassing us.

By 1971, when they presented their work at a momentous conference in Warsaw, Zinaida V. Gordon and Maria N. Sadchikova of the USSR Institute of Labor Hygiene and Occupational Diseases had identified a comprehensive series of symptoms, which they called microwave sickness. Its first signs are low blood pressure and slow pulse. The later and most common manifestations are chronic excitation of the sympathetic nervous system (stress syndrome) and high blood pressure. This phase also often includes headache, dizziness, eye pain, sleeplessness, irritability, anxiety, stomach pain, nervous tension, inability to concentrate, hair loss, plus an increased incidence of appendicitis, cataracts, reproductive problems, and cancer. The chronic symptoms are eventually succeeded by crises of adrenal exhaustion and ischemic heart disease (blockage of coronary arteries and heart attack).

The Soviet standards were set long before the dangers were this clear, however. The comparison is instructive. At a 1969 international symposium on microwaves in Richmond, Virginia, Dr. Karel Marha of Prague's Institute of Industrial Hygiene defended his findings on birth defects and recommended that the Eastern European standard be adopted in the West. Replying to objections that the dire predictions hadn't been proven beyond doubt, he said: "Our standard is not only to prevent damage but to avoid discomfort in people."

Apparently this concern doesn't include Americans, for the Soviets have been bombarding our embassy in Moscow with microwaves for some thirty years. In 1952, at the height of the Cold War, there was a secret meeting at the Sandia Corporation in New Mexico between U.S. and U.S.S.R. scientists, allegedly to exchange information on biological hazards and safety levels. It seems the exchange wasn't completely reciprocal, or perhaps the Americans didn't take seriously what the Russians told them; there have been other joint "workshops" since then, and each time the Soviets have sent people who publicly acknowledged the risks, while the American delegates have always been "no-effect" men. At any rate, soon after the Sandia meeting, the Soviets began beaming microwaves at the U.S. embassy from across Tchaikovsky Street, always staying well within the Schwan limit. In effect, they've been using embassy employees as test subjects for low-level EMR experiments.

The strange thing is that Washington has gone along with it. The "Moscow signal" was apparently first discovered about 1962, when the CIA is known to have sought consultation about it. The agency asked Milton Zaret for information about microwave dangers in that year, and then hired him in 1965 for advice and research in a secret evaluation of the signal, called Project Pandora. Nothing was publicly revealed until 1972, when Jack Anderson broke the story, and the U.S. government told its citizens nothing until 1976, in response to further news stories in the *Boston Globe*. According to various sources, the Russians shut off their transmitter in 1978 or 1979, but then resumed the irradiation for several months in 1983.

According to information given Zaret in the 1960s, the Moscow signal was a composite of several frequencies, apparently aiming for a synergistic effect from various wavelengths, and it was beamed directly at the ambassador's office. Thus it may have been used at least partially to activate bugging devices, but it wasn't consistent with one of the other subsequent official American explanations—a jamming signal to disrupt the U.S. eavesdropping equipment on the embassy roof.

The intensity isn't known for certain. When the State Department admitted the signal's existence, officials claimed it never amounted to more than 18 microwatts. However, although released Project Pandora records don't directly reveal a higher level and the relevant documents have allegedly been destroyed, research protocols aimed at simulating the Moscow signal called for levels up to 4,000 microwatts.

In the mid-1960s published Soviet research indicated that such a beam would produce eyestrain and blurred vision, headaches, and loss of concentration. Within a few years other research had uncovered the entire microwave syndrome, including the cancer potential.

By all accounts except the official ones, the Moscow bombardment has been highly effective. In 1976 the *Globe* reported that Ambassador Walter Stoessel had developed a rare blood disease similar to leukemia and was suffering headaches and bleeding from the eyes. Two of his irradiated predecessors, Charles Bohlen and Llewellyn Thompson, died of cancer. Monkeys exposed to the signal as part of Project Pandora soon showed multiple abnormalities of blood composition and chromosome counts.

In January 1977, the State Department, under duress, announced results of a series of blood tests on returning embassy personnel: a "slightly higher than average" white blood cell count in about a third of the Moscow staff. If 40 percent above the white blood cell counts of other foreign service employees (levels common to incipient leukemia) can be considered "slightly higher than average," then this technically wasn't a lie. The finding has been officially ascribed to some unknown microbe. Unfortunately, there's no such doubt about the veracity of explanations about some earlier research. As part of Project Pandora in the late 1960s, the State Department tested its Moscow employees for genetic damage upon their return stateside, telling them the inner cheek scrapings were to screen for those unusual bacteria. No results were ever released, and they're reportedly part of the missing files, but one of the physicians who conducted the tests was quoted by the Associated Press as saying they'd found "lots of chromosome breaks." The embassy staff had to learn this when the rest of us did—in the newspapers nearly a decade later.

The Russians themselves have never admitted the irradiation, and the Schwan guideline has put the American government in an embarrassing bind. In 1976 the State Department gave its Moscow employees a 20-percent hardship allowance for serving in an "unhealthful post" and installed aluminum window screens to protect the staff from radiation a hundred times weaker than that near many radar bases. That same year the government gave Johns Hopkins School of Medicine a quarter of a million dollars to see if there was a link between the signal and "an apparently high rate of cancer" in the embassy (which wasn't confirmed). Nevertheless, although President Johnson asked Premier Kosygin at the 1967 Glassboro talks to stop the bombardment, Washington has never had any formal basis to demand that it be stopped due to danger to the staff. That was apparently considered an acceptable risk in the protection of the lenient U.S. standard.

Invisible Warfare

The Soviets have led the way in learning about the risks of electropollution, and, as we have seen, they've apparently been the first to harness those dangers for malicious intent. However, the spectrum of potential weapons extends far beyond the limits of the Moscow signal, and Americans have been actively exploring some of them for many years. Most or all of the following EMR effects can be scaled up or down for use against individuals or whole crowds and armies:

> The crudest of these armaments would be a sort of electromagnetic flamethrower with a greater range than chemical types. Dogs were cooked to death in experiments at the Naval Medical Research Institute as long ago as 1955, and high-power transmitters using short UHF wavelengths can severely burn exposed skin in seconds.
>
> Electromagnetic pulse (EMP) is a term designating the immensely powerful, near-instantaneous surge of electromagnetic energy produced by a nuclear explosion. It was first discovered in the late 1960s. The EMP from one detonation a few thousand miles above the earth would destroy all electrical systems throughout an entire continent. In the early 1970s new types of EMR generators emitting power levels ten or twenty times higher than ever before were developed in an effort to simulate EMP and help devise communications systems shielded from it. In 1973 these transmitters were described in an invitation-only seminar at the Naval Weap-

ons Laboratory in Dahlgren, Virginia, where their use for antipersonnel and anti-ballistic-missile energy beams was discussed. No information about their subsequent development has since been made public, and the difficulties of long-range missile tracking argue that ABM beams haven't yet become feasible, but there are no such difficulties in the way of EMR beam weapons for use against unshielded people.

- At some UHF power densities there's an insidious moth-to-the-flame allurement, which would increase such a weapon's effectiveness. As discoverer Sol Michaelson described it in 1958, each of the dogs used in his experiments "began to struggle for release from the sling," showing "considerable agitation and muscular activity," yet "for some reason the animal continues to face the horn." Perhaps as part of the same effect, UHF beams can also induce muscular weakness and lethargy. In Soviet experiments with rats in 1960, five minutes of exposure to 100,000 microwatts reduced swimming time in an endurance test from sixty minutes to six.

- Allen Frey's discovery that certain pulsed microwave beams increased the permeability of the blood-brain barrier could be turned into a supplemental weapon to enhance the effects of drugs, bacteria, or poisons.

- The calcium-outflow windows discovered by Ross Adey could be used to interfere with the functioning of the entire brain.

- In the early 1960s Frey found that when microwaves of 300 to 3,000 megahertz were pulsed at specific rates, humans (even deaf people) could "hear" them. The beam caused a booming, hissing, clicking, or buzzing, depending on the exact frequency and pulse rate, and the sound seemed to come from just behind the head.

At first Frey was ridiculed for this announcement, just like many radar technicians who'd been told they were crazy for hearing certain radar beams. Later work has shown that the microwaves are sensed somewhere in the temporal region just above and slightly in front of the ears. The phenomenon apparently results from pressure waves set up in brain tissue, some of which activate the sound receptors of the inner ear via bone conduction, while others directly stimulate nerve cells in the auditory pathways. Experiments on rats have shown that a strong signal can generate a sound pressure of 120 decibels, or approximately the level near a jet engine at takeoff.

Obviously such a beam could cause humans severe pain and

prevent all voice communication. That the same effect can be used more subtly was demonstrated in 1973 by Dr. Joseph C. Sharp of the Walter Reed Army Institute of Research. Sharp, serving as a test subject himself, heard and understood spoken words delivered to him in an echo-free isolation chamber via a pulsed-microwave audiogram (an analog of the words' sound vibrations) beamed into his brain. Such a device has obvious applications in covert operations designed to drive a target crazy with "voices" or deliver undetectable instructions to a programmed assassin. There are also indications that other pulse frequencies cause similar pressure waves in other tissues, which could disrupt various metabolic processes. A group under R. G. Olsen and J. D. Grissett at the Naval Aerospace Medical Research Laboratory in Pensacola has already demonstrated such effects in simulated muscle tissue and has a continuing contract to find beams effective against human tissues.

In the 1960s Frey also reported that he could speed up, slow down, or stop isolated frog hearts by synchronizing the pulse rate of a microwave beam with the beat of the heart itself. Similar results have been obtained using live frogs, indicating that it's technically feasible to produce heart attacks with a ray designed to penetrate the human chest.

In addition to the methods of damaging or killing people with EMR, there are several ways of controlling their behavior. Ross Adey and his colleagues have shown that microwaves modulated in various ways can force specific electrical patterns upon parts of the brain. Working with cats they found that brain waves appearing with conditioned responses could be selectively enhanced by shaping the microwaves with a rhythmic variation in amplitude (height) corresponding to EEG frequencies. For example, a 3-hertz modulation decreased 10-hertz alpha waves in one part of the animal's brain and reinforced 14-hertz beta waves in another location.

Some radar can find a fly a kilometer away or track a human at twenty-five miles, and several researchers have suggested that focused EMR beams of such accuracy could bend the mind much like electrical stimulation of the brain (ESB) through wires. We know of ESB's potential for mind control largely through the work of José Delgado. One signal provoked a cat to lick its fur, then continue compulsively licking the floor and bars of its cage. A signal designed to stimulate a portion of a monkey's thalamus, a major midbrain center for integrating muscle movements, triggered a complex action: The monkey walked to one side of the cage, then the other, then climbed to the rear ceiling, then back down. The animal performed this same activity as many times as it was stimulated with the signal, up to sixty times an hour, but not blindly—the creature still was able to avoid obstacles and threats from the dominant male while carrying out the electrical imperative. Another type of signal has made monkeys turn their heads, or smile, no matter what else they were doing, up to twenty thousand times in two weeks. As Delgado concluded, "The animals looked like electronic toys."

Even instincts and emotions can be changed: In one test a mother giving continuous care to her baby suddenly pushed the infant away whenever the signal was given. Approach-avoidance conditioning can be achieved for any action simply by stimulating the pleasure and pain centers in an animal's or person's limbic system.

Eventual monitoring of evoked potentials from the EEG, combined with radio-frequency and microwave broadcasts designed to produce specific thoughts or moods, such as compliance and complacency, promises a method of mind control that poses immense danger to all societies—tyranny without terror. Scientists involved in EEG research all say the ability is still years away, but for all we could *sense* of it, it could be happening right now. Conspiracy theories aside, the hypnotic familiarity of TV and radio, combined with the biological effects of their broadcast beams, may already constitute a similar force for mass standardization, whether by design or not.

The potential dangers of televised lethargy are no yawning matter. It's well known that relaxed attention to any mildly involving stimulus, such as a movie or TV program, produces a hypnoid state, in which the mind becomes especially receptive to suggestion. Other inducers of hypnoid states include light sleep, daydreams, or short periods of time spent waiting for some predetermined signal or action, such as a traffic light.

The Central Intelligence Agency funded research on electromagnetic mind control at least as early as 1960, when the notorious MKULTRA program, mostly concerned with hypnosis and psychedelic drugs, included money for adapting bioelectric sensing methods (at that time primarily the EEG) to surveillance and interrogation, as well as for finding "techniques of activation of the human organism by remote electronic means." In testimony before the Senate Subcommittee on Health and Scientific Research on September 21, 1977, MKULTRA director Dr. Sidney Gottlieb recalled: "There was a running interest in what effects people's standing in the field of radio energy have, and it could easily have been that somewhere in the many projects someone was trying to see if you could hypnotize somebody easier if he was standing in a radio beam."

Hypnotists often use a strobe light flashing at alpha-wave frequencies to ease the glide into trance. It seems for over thirty years the Communist bloc nations have been using an ELF wave form to do the same thing undetectably and perhaps more effectively. Ross Adey recently lost most of his government grants and has become a bit more loquacious about the military and intelligence uses of EMR. In 1983 he organized a public meeting at the Loma Linda VA hospital and released photos and information concerning a Russian Lida machine. This was a small transmitter that emitted 10-hertz waves for tranquilization and enhancement of suggestibility. The most interesting part was that the box had an ancient vacuum-tube design, and a man who'd been a POW in Korea reported that similar devices had been used there during interrogation.

American interest in the hypnosis-EMR interaction was still strong as of 1974, when a research plan was filed to develop useful techniques in human volunteers. The experimenter, J. F. Schapitz, stated: "In this investigation it will be shown that the spoken word of the hypnotist may also be conveyed by modulated electromagnetic energy directly into the subconscious parts of the human brain—i.e., without employing any technical devices for receiving or transcoding the messages and without the person exposed to such influence having a chance to control the information input consciously." As a preliminary test of the general concept, Schapitz proposed recording the brain waves induced by specific drugs, then modulating them onto a microwave beam and feeding them back into an undrugged person's brain to see if the same state of consciousness could be produced by the beam alone.

Schapitz's main protocol consisted of four experiments. In the first, subjects would be given a test of a hundred questions, ranging from easy to technical, so they all would know some but not all of the answers. Later, while in hypnoid states and not knowing they were being irradiated, these people would be subjected to information beams suggesting answers for some of the items they'd left blank, amnesia for some of their correct answers, and memory falsification for other correct answers. A new test would check the results two weeks later.

The second experiment was to be the implanting of hypnotic suggestions for simple acts, like leaving the lab to buy some particular item, which were to be triggered by a suggested time, spoken word, or sight. Subjects were to be interviewed later. "It may be expected," Schapitz wrote, "that they rationalize their behavior and consider it to be undertaken out of their own free will."

In a third test the subjects were to be given two personality tests. Then different responses to certain questions would be repeatedly suggested, and nonpathological personality changes would also be suggested, both to be evaluated by new testing in a month. In some cases the subjects were to be prehypnotized into talking in their sleep, so the microwave programmer could gear the commands to thoughts already in the brain. Finally, attempts would be made to produce the standard tests of deep hypnotic trance, such as muscular rigidity, by microwave beams alone.

Naturally, since this information was voluntarily released via the Freedom of Information Act, it must be taken with a pillar of salt. The *results* haven't been made public, so the work may have been inconclusive, and the plans may have been released to convince the Soviets and our own public that American mind-control capabilities are greater than they actually are. On the other hand, the actualities may be so far ahead of this research plan that it was tame enough to release in satisfying FOIA requirements.

How many of the EMR weapons possibilities have actually been developed and/or used? Those not privy to classified information have no way of knowing. There are plenty of rumors. Boris Spassky claimed he'd lost the world chess championship to Bobby Fischer because he was being bombarded with confusion rays. I recall hearing about one secret American experiment in which a scientist was supposedly set up with invitations to three conferences to give the same presentation each time. The first one went fine, but at the last two he was irradiated with ELF waves, reportedly to induce Adey's calcium efflux, and he became confused and ineffective.

Another FOIA release from the Defense Intelligence Agency in 1976 may be revealing. Prepared by Ronald L. Adams and E. A. Williams of Battelle Columbus Laboratories, it's entitled "Biological Effects of Electromagnetic Radiation (Radiowaves and Microwaves), Eurasian Communist Countries." The pages released merely recount Allen Frey's discoveries without mentioning his name, implying instead that only the Reds would be so dastardly as to investigate such things for use as weapons. Immediately after mention of the blood-brain barrier leak phenomenon, a paragraph was deleted, followed by the tantalizing sentence, "The above study is recommended reading material for those consumers who have an interest in the application of microwave energy to weapons." Even without this document, considering the relentless pace of arms development, we would have to be very naïve to assume that the United States has no electromagnetic arsenal.

The Soviets may already be using theirs, however, on a scale far beyond that of the Moscow signal. During the U.S. bicentennial celebration of July 4, 1976, a new radio signal was heard throughout the world. It has remained on the air more or less continuously ever since. Varying up and down through the frequencies between 3.26 and 17.54 megahertz, it is pulse-modulated at a rate of several times a second, so it sounds like a buzz saw or woodpecker. It was soon traced to an enormous transmitter near Kiev in the Soviet Ukraine.

The signal is so strong it drowns out anything else on its wavelength. When it first appeared, the UN International Telecommunications Union protested because it interfered with several communications channels, including the emergency frequencies for aircraft on transoceanic flights. Now the woodpecker leaves "holes"; it skips the crucial frequencies as it moves up and down the spectrum. The signal is maintained at enormous expense from a current total of seven stations, the seven most powerful radio transmitters in the world.

Within a year or two after the woodpecker began tapping, there were persistent complaints of unaccountable symptoms from people in several cities of the United States and Canada, primarily Eugene, Oregon. The sensations—pressure and pain in the head, anxiety, fatigue, insomnia, lack of coordination, and numbness, accompanied by a high-pitched ringing in the ears—were characteristic of strong radio-frequency or microwave irradiation. In Oregon, between Eugene and Corvallis, a powerful radio signal centering on 4.75 megahertz was monitored, at higher levels in the air than on the ground. Several unsatisfactory theories were advanced, including emanations from winter-damaged power lines, but most engineers who studied the signal concluded that it was a manifestation of the woodpecker. The idea was advanced that it was being directed to Oregon by a Tesla magnifying transmitter. This apparatus, devised by Nikola Tesla during his turn-of-the-century experiments on wireless global power transmission at a laboratory near Pikes Peak, hasn't been much studied in the West. It reportedly enables a transmitter to beam a radio signal *through* the earth to any desired point on its surface, while maintaining or even increasing the signal's power as it emerges. Paul Brodeur has suggested that, since the TRW company once proposed a Navy ELF communications system using an existing 850-mile power line that ended in Oregon, the Eugene phenomenon might have been the interaction between a Navy broadcast and Soviet jamming.

Be that as it may, the woodpecker continues in operation, and there are several unsettling possibilities as to its main purpose. A former chief of naval research has privately discounted the idea that it's directed against the U.S. population. However, Robert Beck, a Los Angeles physicist who regularly serves as a DOD consultant, told me that the signal has a threefold purpose. He said it acts as a crude over-the-horizon radar that would pick up a massive first strike of U.S. missiles if Soviet spy satellites and other detectors were knocked out. Second, the signal's modulations are an ELF medium for communicating with submarines underwater. Third, he claimed the signal has a biological by-product about which he promised further information. Of course, I haven't been able to contact him since.

Several educated guesses can be made, however. Adey's research suggests that the best way to get an ELF signal into an animal is to make it a pulse modulation of a high-frequency radio signal. That's exactly what the woodpecker is. Within its frequency range, it could be beamed to any part of the world, and it would be picked up and reradiated by the power supply grid at its destination.

Raymond Damadian has theorized that the woodpecker signal is designed to induce nuclear magnetic resonance in human tissues. Damadian, a radiologist at Brooklyn's Downstate Medical Center, patented the first NMR scanner, a device that gives an image of internal organs similar to CAT scanners but using magnetic fields rather than nuclear radiation. As mentioned earlier in this chapter, NMR could greatly magnify the metabolic interference of electropollution or EMR weapons. Maria Reichmanis calculated the pulse frequency that would be required to do this with a radio signal in the woodpecker's range, and she came up with a band centered on the same old alpha rhythm of 10 hertz. And in fact, the signal's pulse is generally about that rate, although it is often a two-part modulation of $4 + 6$, $7 + 3$, and so on. The available evidence, then, suggests that the Russian woodpecker is a multipurpose radiation that combines a submarine link with an experimental attack on the American people. It may be intended to increase cancer rates, interfere with decision-making ability, and/or sow confusion and irritation. It may be succeeding.

I keep hearing persistent rumors of American transmitters set up to try to nullify the Russians' signal or to affect their people in a similar way. In 1978, Stefan Rednip, an American reporter living in England, claimed access to purloined CIA documents proving the existence of a program called Operation Pique, which included bouncing radio signals off the ionosphere to affect the mental functions of people in selected areas, including Eastern European nuclear installations.

The whole business sounds too much like an undeclared electromagnetic war. However, there are persistent complaints that the American effort is being hampered in a strange way. Shortly after the rigged National Academy of Sciences report on Project Seafarer, for example, the Navy sent a delegation to a meeting at the National Security Agency to complain about an alleged "zap gap" between the United States and the USSR, and to ask other delegates to push for more research money for turning nonthermal EMR effects into weapons. According to one of my Navy contacts, the NSA sent several "experts" who had never done any research on EMR and who firmly advised the Navy to abandon its program. Later he voiced the same suspicions I'd already heard from others: Given the allegedly vigorous Soviet electroweapons research program and the underfunding of ours, he concluded that there is a mole highly placed in the American military science establishment, perhaps in the NSA itself, who is preventing us from acquiring any clear competence in this field.

Unfortunately, my source, having served as a hatchet man for defunding research on the environmental dangers of electropollution, isn't exactly reliable. Complaints of a mole could easily be a blind for a large and intense U.S. EMR weapons program. That there's more going on than meets the eye is clear from my last communication with Dietrich Beischer. In 1977 the Erie Magnetics Company of Buffalo, New York, sponsored a small private conference, and Beischer and I both planned to attend. Just before the meeting, I got a call from him. With no preamble or explanation, he blurted out: "I'm at a pay phone. I can't talk long. They are watching me. I can't come to the meeting or ever communicate with you again. I'm sorry. You've been a good friend. Goodbye." Soon afterward I called his office at Pensacola and was told, "I'm sorry, there is no one here by that name," just as in the movies. A guy who had done important research there for decades just disappeared.

The crucial point to me is that both sides may be embarking on hostilities whose consequences for the whole biosphere no one can yet foresee. Even if the Soviets have begun an electromagnetic war and we're totally unprepared to fight back, I doubt that a simple buildup and retaliation are the best course for our own survival.

The extent of the danger can be dramatized best by considering one last potential weapon. Around 1900, Nikola Tesla theorized that ELF and VLF radiation could enter the magnetosphere, the magnetic field in space around the earth, and change its structure. He has recently been proven right.

The magnetosphere and its Van Allen belts of trapped particles produce many kinds of EMR. Since they were initially studied through audio amplifiers, the first kinds to be discovered, around 1920, were given fanciful names like whistlers, dawn chorus, and lion roars. Many of them result from VLF waves produced by lightning, which bounce back and forth from pole to pole along "magnetic ducts" in the magnetosphere. This resonance amplifies the original VLF waves enormously.

Satellite measurements have proven that artificial energies from power lines are similarly amplified high above the earth, a phenomenon known as power-line harmonic resonance (PLHR). Radio and microwave energy also resonates in the magnetosphere. This amplified energy interacts with the particles in the Van Allen belts, producing heat, light, X rays, and, most important, a "fallout" of charged particles that serve as nuclei for raindrops.

Recent work with sounding rockets has matched specific areas of such ion precipitation with the energy from specific radio stations, and established that the sifting down of charged particles generally occurs east of the EMR source, following the general eastward drift of weather patterns. In 1983, measurements from the Ariel 3 and 4 weather satellites showed that the enormous amount of PLHR over North America had created a permanent duct from the magnetosphere down into the upper air, resulting in a continuous release of ions and energy over the whole continent. In presenting this data at the March 1983 Symposium on Electromagnetic Compatibility in Zurich, K. Bullough reminded the audience that thunderstorms have been 25 percent more frequent over North America between 1930 and 1975 than they were from 1900 to 1930, and suggested that the increased energy levels in the upper atmosphere were responsible.

Since the mid-1970s there has been a dramatic increase in flooding, drought, and attendant hardships due to inconsistent, anomalous weather patterns. It appears likely that these have been caused in part by electropollution and perhaps enhanced, whether deliberately or not, by the Soviet woodpecker signal. It now seems feasible to induce catastrophic climate change over a target country, and even without such weather warfare, continued expansion of the electrical power system threatens the viability of all life on earth.

Critical Connections

It may be hard to convince ourselves that something we can't see, hear, touch, taste, or smell can still hurt us so dreadfully. Yet the fact must be faced, just as we've learned a healthy fear of nuclear radiation. Certain scientists, some perhaps acting in a program of deliberate disinformation, keep telling the public that we still don't know whether electropollution is a threat to human health. That's simply not true. Certainly we need to know more, but a multitude of risks have been well documented.

Three dangers overshadow all others. The first has been conclusively proven: *ELF electromagnetic fields vibrating at about 30 to 100 hertz, even if they're weaker than the earth's field, interfere with the cues that keep our biological cycles properly timed; chronic stress and impaired disease resistance result.* Second, the available evidence strongly suggests that regulation of cellular growth processes is impaired by electropollution, increasing cancer rates and producing serious reproductive problems. Electromagnetic weapons constitute a third class of hazards culminating in climatic manipulation from a sorcerer's-apprentice level of ignorance.

There may be other dangers, less sharply defined but no less real. All cities, by their very nature as electrical centers, are jungles of interpenetrating fields and radiation that completely drown out the earth's background throb. Is this an underlying reason why so many of them have become jungles in another sense as well? Is this a partial explanation for the fact that the rate of suicide between the ages of fifteen and twenty-four rose from 5.1 per 100,000 in 1961 to 12.8 in 1981? Might this be an invisible and thus overlooked reason why so many governmental leaders, working at the centers of the most powerful electromagnetic networks, consistently make decisions that are against the best interests of every being on earth? Is the subliminal stress of electronic smog misinterpreted as continual threats from outside—from other people and other governments? In addition, if Teilhard de Chardin's noosphere exists, our artificial fields must mask it many times over, literally disconnecting us from life's collective wisdom. This is not to ignore the plain fact of evil, but it often seems there must be some other reason why today's power elite are so willing to bring the whole world to the brink of so many different kinds of destruction. Maybe they literally can't hear the earth anymore.

Everyone worries about nuclear weapons as the most serious threat to our survival. Their danger is indeed immediate and overwhelming. In the long run, however, I believe the ultimate weapon is manipulation of our electromagnetic environment, because it's imperceptibly subtle and strikes at the core of life itself. We're dealing here with the most important scientific discovery *ever*—the nature of life. Even if we survive the chemical and atomic threats to our existence, there's a strong possibility that increasing electropollution could set in motion irreversible changes leading to our extinction before we're even aware of them.

All life pulsates in time to the earth, and our artificial fields cause abnormal reactions in all organisms. Magnetic reversals may have produced the "great dyings" of the past by disrupting biocycles so as to cause stress, sterility, birth defects, malignancies, and impaired brain function. Human activities may well have duplicated in three decades what otherwise would have taken five thousand years to develop during the next reversal. What will we do if the incidence of deformed children rises to 50 percent, if the cancer rate climbs to 75 percent? Will we be able to pull the plug?

Somehow these dangers must be brought into the open so forcefully that the entire population of the world is made aware of them. Scientists must begin to ask and seek answers to the questions raised in this chapter, regardless of the effect on their careers. These energies are too dangerous to be entrusted forever to politicians, military leaders, and their lapdog researchers.

Since our civilization is irreversibly dependent on electronics, abolition of EMR is out of the question. However, as a first step toward averting disaster, we must halt the introduction of new sources of electromagnetic energy while we investigate the biohazards of those we already have with a completeness and honesty that have so far been in short supply. New sources must be allowed only after their risks have been evaluated on the basis of the knowledge acquired in such a moratorium.

With an adequately funded research program, the moratorium need last no more than five years, and the ensuing changes could almost certainly be performed without major economic trauma. It seems possible that a different power frequency—say 400 hertz instead of 60—might prove much safer. Burying power lines and providing them with grounded shields would reduce the electric fields around them, and magnetic shielding is also feasible.

A major part of the safety changes would consist of energy-efficiency reforms that would benefit the economy in the long run. These new directions would have been taken years ago but for the opposition of power companies concerned with their short-term profits, and a government unwilling to challenge them. It is possible to redesign many appliances and communications devices so they use far less energy. The entire power supply could be decentralized by feeding electricity from renewable sources (wind, flowing water, sunlight, geothermal and ocean thermal energy conversion, and so forth) into local distribution nets. This would greatly decrease the hazards by reducing the voltages and amperages required. Ultimately, most EMR hazards could be eliminated by the development of efficient photoelectric converters to be used as the primary power source at each point of consumption. The changeover would even pay for itself, as the loss factors of long-distance power transmission—not to mention the astronomical costs of building and decommissioning short-lived nuclear power plants—were eliminated. Safety need not imply giving up our beneficial machines.

Obviously, given the present technomilitary control of society in most parts of the world, such sane efficiency will be immensely difficult to achieve. Nevertheless, we must try. Electromagnetic energy presents us with the same imperative as nuclear energy: Our survival depends on the ability of upright scientists and other people of goodwill to break the military-industrial death grip on our policy-making institutions.

Postscript: Political Science

An important scientific innovation rarely makes its way by
gradually winning over and converting its opponents: it
rarely happens that Saul becomes Paul. What does happen
is that its opponents gradually die out and that the
growing generation is familiarized with the idea from the
beginning.

—MAX PLANCK

Dispassionate philosopher inquiring into nature from the sheer love of
knowledge, single-minded alchemist puttering about a secluded base-
ment in search of elixirs to benefit all humanity—these ideals no longer
fit most scientists. Even the stereotype of Faust dreaming of demonic
power is outdated, for most scientists today are overspecialized and
anonymous—although science as a whole is somewhat Mephistophelian
in its disregard for the effects of its knowledge. It's a ponderous beast,
making enormous changes in the way we live but agonizingly slow to
change its own habits and viewpoints when they become outmoded.

The public's conception of the scientist remains closest to its image of
the philosopher—cold and logical, making decisions solely on the basis
of the facts, unswayed by emotion. The lay person's most common fear
about scientists is that they lack human feelings. During my twenty-five

years of research I've found this to be untrue yet no cause for comfort. I've occasionally seen our species' nobler impulses among them, but I've also found that scientists as a group are at least as subject to human failings as people in other walks of life.

It has been like this throughout the history of science. Many, perhaps even most, of its practitioners have been greedy, power-hungry, prestige-seeking, dogmatic, pompous asses, not above political chicanery and outright lying, cheating, and stealing. Examples abound right from the start. Sir Francis Bacon, who in 1620 formulated the experimental method on which all technical progress since then has been founded, not only forgot to mention his considerable debt to William Gilbert but apparently plagiarized some of his predecessor's work while publicly belittling it. In a similar way Emil Du Bois-Reymond based his own electrical theory of the nerve impulse on Carlo Matteucci's work, then tried to ridicule his mentor and take full credit.

Many a genius has been destroyed by people of lesser talent defending the status quo. Ignaz Semmelweis, a Hungarian physician who practiced in Vienna during the mid-nineteenth century, demanded that his hospital colleagues and subordinates wash their hands, especially when moving from autopsies and sick wards to the charity childbirth ward he directed. When the incidence of puerperal fever and resultant death declined dramatically to well below that of the rich women's childbirth ward, proving the importance of cleanliness even before Pasteur, Semmelweis was fired and vilified. His livelihood gone, he committed suicide soon afterward.

The principal figure who for decades upheld the creed that dedifferentiation was impossible was Paul Weiss, who dominated biology saying the things his peers wanted to hear. Weiss was wrong, but along the way he managed to cut short a number of careers.

For many years the American Medical Association scorned the idea of vitamin-deficiency diseases and called the EEG electronic quackery. Even today that august body contends that nutrition is basically irrelevant to health. As the late-eighteenth-century Italian experimenter Abbé Alberto Fortis observed in a letter chiding Spallanzani for his closed-minded stance on dowsing, ". . . derision will never help in the development of true knowledge."

In the past, these character flaws couldn't wholly prevent the recognition of scientific truths. Both sides of a controversy would fight with equal vehemence, and the one with better evidence would usually win sooner or later. In the last four decades, however, changes in the structure of scientific institutions have produced a situation so heavily weighted in favor of the establishment that it impedes progress in health care and prevents truly new ideas from getting a fair hearing in almost all circumstances. The present system is in effect a dogmatic religion with a self-perpetuating priesthood dedicated only to preserving the current orthodoxies. The system rewards the sycophant and punishes the visionary to a degree unparalleled in the four-hundred-year history of modern science.

This situation has come about because research is now so expensive that only governments and multinational corporations can pay for it. The funds are dispensed by agencies staffed and run by bureaucrats who aren't scientists themselves. As this system developed after World War II, the question naturally arose as to how these scientifically ignorant officials were to choose among competing grant applications. The logical solution was to set up panels of scientists to evaluate requests in their fields and then advise the bureaucrats.

This method is based on the naïve assumption that scientists really *are* more impartial than other people, so the result could have been predicted decades ago. In general, projects that propose a search for evidence in support of new ideas aren't funded. Most review committees approve nothing that would challenge the findings their members made when they were struggling young researchers who created the current theories, whereas projects that pander to these elder egos receive lavish support. Eventually those who play the game become the new members of the peer group, and thus the system perpetuates itself. As Erwin Chargaff has remarked, "This continual turning off and on of the financial faucets produces Pavlovian effects," and most research becomes mere water treading aimed at getting paid rather than finding anything new. The intuitive "lunatic twinge," the urge to test a hunch, which is the source of all scientific breakthroughs, is systematically excluded.

There has even been a scientific study documenting how choices made by the peer review system depend almost entirely on whether the experts are sympathetic or hostile to the hypothesis being suggested. True to form, the National Academy of Sciences, which sponsored the investigation, suppressed its results for two years.

Membership on even a few peer review boards soon establishes one's status in the "old boys' club" and leads to other benefits. Manuscripts submitted to scientific journals are reviewed for validity in the same way as grant requests. And who is better qualified to judge an article than those same eminent experts with their laurels to guard? Publication is accepted as evidence that an experiment has some basic value, and without it the work sinks without a ripple. The circle is thus closed, and the revolutionary, from whose ideas all new scientific concepts come, is on the outside. Donald Goodwin, chairman of psychiatry at the University of Kansas and an innovative researcher on alcoholism, has even put it in the form of a law of exasperation: "If it's trivial, you can probably study it. If it's important, you probably can't."

Another unforeseen abuse has arisen, which has lowered the quality of training in medical schools. As the peer review system developed, academic institutions saw a golden opportunity. If the government wanted all this research done, why shouldn't it help the schools with their overhead, such as housing, utilities, bookkeeping, and ultimately the salaries of the researchers, who were part of the faculty? The influx of money corroded academic values. The idea arose that the best teacher was the best researcher, and the best researcher was the one who pulled down the biggest grants. A medical school became primarily a kennel of researchers and only secondarily a place to teach future physicians. To survive in academia, you have to get funded and then get published. The epidemic of fraudulent reports—and I believe only a small percentage of the actual fakery has been discovered—is eloquent testimony of the pressure to make a name in the lab.

There remain today few places for those whose talents lie in teaching and clinical work. Many people who don't care about research are forced to do it anyway. As a result, medical journals and teaching staffs are both drowning in mediocrity.

Finally, we must add to these factors the buying of science by the military. To call it a form of prostitution is an insult to the oldest profession. Nearly two thirds of the $47-billion 1984 federal research budget went for military work, and in the field of bioelectricity the proportion was even higher. While military sponsors often allow more technical innovation than others, their employees must keep their mouths shut about environmental hazards and other moral issues that link science to the broader concerns of civilization. In the long run, even the growth of pure knowledge (if there is such a thing) can't flourish behind this chain link fence.

If someone does start a heretical project, there are several ways of dealing with the threat. Grants are limited, usually for a period of one or two years. The experimenter then must reapply. Every application is a voluminous document filled with fine-print forms and meaningless bureaucratic jargon, requiring many days of data compilation and "creative writing." Some researchers may simply get tired of them and quit. In any case, they must run the same gauntlet of peers each time. The simplest way to nip a challenge in the bud is to turn off the money or keep the reports out of major journals by means of anonymous value judgments from the review committees. You can *always* find something wrong with a proposal or manuscript, no matter how well written or scientifically impeccable it may be.

Determined rebels use guerrilla tactics. There are so many funding agencies that the left hand often knoweth not what the right hand doeth. A proposal may get by an obscure panel whose members aren't yet aware of the danger. The snowstorms of paper churned out by the research establishment have required the founding of many new journals in each subspecialty. Some of these will accept papers that would automatically be rejected by the big ones. In addition, there's an art to writing a grant proposal that falls within accepted guidelines without specifying exactly what the researchers intend to do.

If these methods succeed in prolonging the apostasy, the establishment generally exerts pressure through the schools. Successful academics are almost always true believers who are happy to curry favor by helping to deny tenure to "questionable" investigators or by harassing them in a number of ways. For example, in 1950 Gordon A. Atwater was fired as chairman of the American Museum of Natural History astronomy department and curator of the Hayden Planetarium for publicly suggesting that Immanuel Velikovsky's ideas should receive a fair hearing. That same year Velikovsky's first book, *Worlds in Collision,* was renounced by his publisher (Macmillan) even though it was a best seller, because a group of influential astronomers led by Harvard's Harlow Shapley threatened to boycott the textbook department that accounted for two thirds of the company's sales. No matter what one may think of Velikovsky's conclusions, that kind of backstairs persuasion is not science.

As the conflict escalates, the muzzled freethinker often goes directly to the public to spread the pernicious doctrines. At this point the gloves come off. Already a lightning rod for the wrath of the Olympian peers, the would-be Prometheus writhes under attacks on his or her honesty, scientific competence, and personal habits. The pigeons of Zeus cover the new ideas with their droppings and conduct rigged experiments to disprove them. In extreme cases, government agencies staffed and advised by the establishment begin legal harassment, such as the trial and imprisonment that ended the career and life of Wilhelm Reich.

Sometime during or after the battle, it generally becomes obvious that the iconoclast was right. The counterattack then shifts toward historical revision. Establishment members publish papers claiming the new ideas for themselves and omitting all references to the true originator. The heretic's name is remembered only in connection with a condescending catchphrase, while his or her own research programs, if any remain, are defunded and the staff dispersed. The facts of the case eventually emerge, but only at an immense toll on the innovator's time and energy.

To those who haven't tried to run a lab, these may seem like harsh words, unbelievable, even paranoid. Nevertheless, these tactics are commonplace, and I've had personal experience with each and every one of them.

I got a taste of the real world in my very first foray into research. After World War II, I continued my education on the GI Bill, but those benefits expired in 1947. I'd just married a fellow student named Lillian, who had caught my eye during our first orientation lecture, and I needed a summer job to help pay expenses and set up housekeeping. I was lucky enough to get work as a lab assistant in the NYU School of Medicine's surgical research department.

I worked with Co Tui, who was evaluating a recently published method for separating individual amino acids from proteins as a step toward concentrating foods for shipment to the starving. Dr. Co, a tiny man whose black, spiky hair seemed to broadcast enthusiasm, inspired me enormously. He was a brilliant researcher and a good friend. With him I helped develop the assay technique and began to use it to study changes in body proteins after surgery.

I was writing my first scientific paper when I walked to work one morning and found our laboratory on the sidewalk—all our equipment, notes, and materials junked in a big pile. I was told neither of us worked there anymore; we were welcome to salvage anything we wanted from the heap.

The head secretary told me what had happened. This was during a big fund drive to build the present NYU Medical Center. One of the "society surgeons" had lined up a million-dollar donation from one of his patients and would see that it got into the fund, if he could choose a new professor of experimental surgery—*now*. As fast as that, Co Tui and his people were out. I vowed to Lillian: "Whatever I do in medicine, I'm going to stay out of research."

I'm happy that I wasn't able to keep my promise. The research itself was worth it all. Moreover, I don't want to give the impression that I and my associates were alone against the world. Just when hope seemed lost, there was always a crucial person, like Carlyle Jacobsen or the research director's secretary, to help us out. However, right from my first proposal to measure the current of injury in salamanders, I found that research would mean a constant battle, and not only with administrators.

Before I began, I had to solve a technical problem with the electrodes. Even two wires of the same metal had little chemical differences, which gave rise to small electrical currents that could be misinterpreted as coming from the animal. Also, the slightest pressure on the animal's skin produced currents. No one understood why, but there they were. I found descriptions in the older literature of silver electrodes with a layer of silver chloride applied to them, which were reported to obviate the false interelectrode currents. I made some, tested them, and then fitted them with a short length of soft cotton wick, which got rid of the pressure artifact. When I wrote up my results, I briefly described the electrodes. Afterward I received a call from a prominent neurophysiologist who wanted to visit the lab. "Very nice," I thought. "Here's some recognition already." He was particularly interested in how the electrodes were made and used. Some months later, damned if I didn't find a paper by my visitor in one of the high-class journals, describing this new and excellent electrode he'd devised for measuring direct-current potentials!

A couple of years later, while Charlie Bachman and I were looking for the PN junction diode in bone, I was asked to give a talk on bone electronics at a meeting in New York City. The audience included engineers, physicists, physicians, and biologists. It was hard to talk to such a diverse group. The engineers and physicists knew all about electronics but nothing about bone, the biologists knew all about bone but nothing about electronics, and the physicians were only interested in therapeutic applications. At any rate, I reviewed some bone structure for the physicists and some electronics for the biologists, and then went on to describe my experiments with Andy Bassett on bone piezoelectricity.

I probably should have sat down at that point, but I thought it would be nice to talk about our present work. The rectifier concept was tremendously exciting to me, and I thought we might get some useful suggestions from the audience, so I described the experiments showing that collagen and apatite were semiconductors, and discussed the implications. After each talk a short time was set aside for questions and comments, generally polite and dignified. However, as soon as I finished, a well-known orthopedic researcher literally ran up to the audience microphone and blurted out, "I have never heard such a collection of inadequate data and misconceptions. It is an insult to this audience. Dr. Becker has not presented satisfactory evidence for any semiconducting property in bone. The best that can be said is that this material may be a semi-insulator."

Semiconductors are so named because their properties place them between conductors and insulators, so you could very well call them semi-insulators; the meaning would be the same. My opponent was playing a crude game. While saying these derogatory things about me, he was actually agreeing with my conclusion, merely using a different term.

This man's antagonism had begun a couple of years before. When Andy Bassett and I had finished our work on the piezoelectric effect in bone, we wrote it up, submitted it to a scientific journal, and got it accepted. Unbeknownst to us, this fellow had been working on the same thing, but hadn't gotten as far in his experiments as we. Somehow he learned of our work and its impending publication. He called Andy, asking us to delay our report until he was ready to publish his own data. Andy called me to talk it over. What counts in the scientific literature is *priority;* he was asking us to surrender it. There was no ethical basis for his request, and I would never have thought of asking him to delay had the situation been reversed. I said, "Not on your life." Our paper was published, and we'd acquired a "friend" for life.

Now there he was at the microphone trying to scuttle my presentation with a little ambiguous double-talk. I thought, "He must be doing the same work as we are again. If he wins this encounter, I'll have trouble getting my data published, and he'll have a clear field for his." Instead of defending the data, I explained that semi-insulator and semiconductor were one and the same. I said I was surprised he didn't know that, but I appreciated his approval of my data! Someone else in the audience stood up in support of my position, and the crisis was past. The lab isn't the only place a scientist has to stay alert.

In 1964, soon after the National Institutes of Health approved the grant for our continuing work on bone, I received the VA's William S. Middleton Award for outstanding research. That's a funny little story in itself. The award is given by the VA's Central Office (VACO), whose members had already decided on me, but candidates must be nominated by regional officers, and the local powers were determined I shouldn't get it. Eventually VACO had to *order* them to nominate me.

The award put me on salary from Washington instead of Syracuse, and due to pressure from VACO I was soon designated the local chief of research, replacing the man who signed all the papers at once. I was determined to put the research house in order, and I instituted a number of reforms, such as public disclosure of the funding allocations, and productivity requirements, no matter how prominent an investigator might be. Many of the reforms have been adopted throughout the VA system. They didn't make me more popular, however. Over the next several years there was continuous pressure from the medical school to allocate

VA research funds for people I felt were of little value to the VA program itself; thus the money would have constituted a grant to the school. I knew that if I didn't deliver I would eventually be removed from the position of chief of research. In that case, I would go back on a local clinical salary and my research program would again be in jeopardy. Therefore, at the beginning of 1972 I applied for the position of medical investigator in the VA research system, a post in which I would be able to devote up to three fourths of my time to research. I was accepted. The job was to begin a few months later; in the meantime I continued as chief of research.

Apparently my new appointment escaped the notice of my local opponents. I'd accepted several invitations to speak at universities in the South and combined them all into a week's trip. I left the office a day early to prepare my material and pack. While I was still home, my secretary called. She was crying, and said she'd just gotten a memo firing me as chief of research and putting me to work as a general-duty medical officer in the admitting office. This not only would have closed our lab, but also would have kept me from practicing orthopedic surgery.

It was a nice maneuver but, fortunately for me, it wasn't legal. As medical investigator, I could be fired only by Washington, and the local chief of staff soon got a letter from VACO ordering him to reinstate me.

Soon I began to get on some "enemies lists" at the national level, too. In December 1974 I got word that our basic NIH grant (the one on bone) hadn't been renewed. No reasons were given. This was highly irregular, since applicants normally got the "pink sheets" with at least the primary reviewer's comments, so they could find out what they'd done wrong. Instead I was told I could write to the executive secretary for a "summary" of the deliberations.

The summary was half a page of double-spaced typing. It said my proposal had been lacking in clarity and direction, and that the experimental procedures hadn't been spelled out in enough detail. The main problem seemed to be that I was planning to do more than the reviewer thought I could do with the money I was requesting. In addition, my report on the perineural cell research with Bruce Baker was criticized as "data poor." The statement concluded: "On the other hand, there are some areas which appear to be worthy of support and are reasonably well described, e.g., bone growth studies, regenerative growth, and electrical field effects."

I was, to say the least, puzzled. The subjects "worthy of support" were precisely the main ones we were working on. It didn't make any sense until I reflected that this was just after I'd helped write the first Sanguine report and had begun to testify about power line dangers before the New York Public Service Commission. Perhaps the Navy was pressuring the NIH to shut me up.

If somebody at the federal level was trying to lock me out as early as 1974, he forgot to watch all the entrances, for my proposal of that year on acupuncture was approved. I'd originally tacked this on to the main NIH application, where it was criticized as inappropriate. I merely sent it off to a different study section, which funded it. After a year we had the positive results described in Chapter 13, and I presented them at an NIH acupuncture conference in Bethesda, Maryland. Ours was the only study going at the problem from a strictly scientific point of view, that is, proceeding from a testable hypothesis, as opposed to the empirical approach of actually putting the needles in and trying to decide if they worked. To the NIH's basic question—is the system of points and lines real?—our program was the only one giving an unequivocal answer: yes.

Nevertheless, when this grant came up for renewal in 1976, it, too, was cut off. The stated reasons were that we hadn't published enough and that the electrical system we found didn't have any relation to acupuncture. The first was obviously untrue—we'd published three papers, had two more in press, and had submitted six others—and the second was obvious pettifogging. How could anyone know what was related to acupuncture before the research had been done? I happened to know the chairman of the NIH acupuncture study section, so I wrote him a letter. He said he was surprised, because the group itself had been pleased with our report. By then it was obvious that something was up.

As of October 1976 we would have no more NIH support. As the money dwindled, we juggled budgets and shaved expenses to cover our costs, and with the help of Dave Murray, who was now chairman of the orthopedic surgery department at the medical school, we kept the laboratory intact and enormously productive. We actually published more research than when we hadn't been under fire.

Early in that same year, however, my appointment as medical investigator had expired, and I had to reapply. Word came back that my application was being "deferred," that is, it had been rejected, but I had the option of reapplying immediately. In her accompanying letter, the director of the VA's Medical Research Service wrote, "While your past record and the strong letters of support [the peer reviews of my application] were considered extremely favorable, the broad research proposal with sketchy detail of technique and methodology was not considered approvable." Now, the instructions for medical investigator applications clearly stated that I was to spell out *past* accomplishments and indicate future directions only in broad outline. Instead, the director was applying the criteria for first-time grant applicants just entering research. She invited me to resubmit the proposal in the other format. But that would not have helped. Even if the second application was approved, the money would arrive six months after the lab had been closed and we had gone our separate ways.

There was another strange thing about the rejection. By that time all federal granting agencies *had* to provide the actual reports (with names deleted) of the peers who had done the reviewing. Three out of the four were long, detailed, well-thought-out documents in the standard critique format; they'd been neatly retyped, single spaced, on "reviewer's report" forms with an elite typewriter. One was absolutely lavish in its praise, saying that the VA was fortunate to have me and that the proposed work would undoubtedly make great contributions to medicine. Another was almost as laudatory.

One name had inadvertently been left on one page of the third review. It was the name of a prominent orthopedic researcher with whom I'd disagreed for years about commercialization of bone-healing devices. Since our mutual disregard was well known in the orthopedic service, I feel it was indefensible for the director to ask him to review my application in the first place. Perhaps she expected a more damaging critique from him. He did complain that the proposal was insufficiently detailed. However, his appraisal was quite fair and even said my proposed work was of "fundamental importance to the field of growth and healing." It obviously led up to a recommendation for approval, but the last sentence of that paragraph had been deleted.

The last review was half a page of vague objections, typed double spaced on a pica machine with no semblance of the standard format. There was a revealing mistake ("corrective" tissue instead of connective tissue) that showed the writer had glanced at my proposal for cues but really didn't know what it was about. Strangest of all was the phrasing of this pseudoreview: "[Becker's proposal] is broad and sweeping in scope and contains little documentation for technique and methodology. However, in view of his past record and strong letters of support, a decision should be deferred. . . ." The director had used it almost word for word in her letter.

She certainly had no motive for such conduct herself. I'd met her briefly a few years before. In 1966 she'd been appointed chief of research at the Buffalo VA Medical Center and had visited Syracuse to see how I'd organized the program there. Our conversation was pleasant but quite innocuous.

The incentive came from elsewhere. In 1982 a former chief of naval research, with whom I'd made friends because of our common interest in spinal cord regeneration, told me the Navy had protested the closing of the lab. He said it had been a deliberate action coming from some level higher than the VA, NIH, or NAS. Somewhere the decision was made to "shut this guy up," he said, but he wanted me to know that he and the Navy had had no part in it. At one point during the struggle, Andy Marino got a call from a minor official in the VA. He said, "Listen, we're under an awful lot of pressure here to shut you fellows off. Can't you just back off a little bit, stop saying all these things, just sort of downgrade that public hearing you're involved with?" This was the reasonable approach—"We're only trying to help you"—so Andy spent some time sympathizing with this guy's position. Then when he finally asked him, "Where are you getting the pressure from?" Andy got an answer: "Mostly from DOD."

At about the same time I received a clear message about what was going on from one of the scientists from the "other side," one of those who were doing a lot of well-funded work for the military but publishing little. During a recess in a scientific meeting, I encountered him in an empty lobby. Glancing over his shoulder, he drew me aside to the windows and told me my only trouble was that I was going public. He said he and all the rest of us knew there were nonthermal effects and hazards, but we had to keep it quiet. I replied that if no one "went public," the situation could never be corrected and a lot of people would suffer needlessly. He told me that was no concern of mine and predicted my attitude would ruin my career. Well, I could only agree with him on the last point. I agreed that it probably already had, but at least I had a clear conscience.

Rather than sit and wait for the executioner, I tried the back door. I sent a detailed merit-review proposal, the kind filed by beginning researchers, to a newly established section on rehabilitation research. Instead of applying through VACO, however, I routed it through the VA regional office in Boston.

While I was waiting for a reply, I had to fend off another attack. Before testifying at the PSC hearings, I'd notified my supervisor. With a VA lawyer present, he'd told me in person that this was just the kind of thing the VA wanted to fulfill its public obligations, and he even sent me a letter to that effect. Then, just after I appeared on *60 Minutes* and talked about Sanguine, power lines, and the rigging of the NAS committee, I heard the VA Personnel Office was investigating me for engaging in nonapproved work—the testimony—on government time. I'd gone to the hearings at my own expense to attack a *public* health problem, but never mind; they were still accusing me of stealing from the taxpayers. If the charge was upheld, I would have to pay back part of my salary and grants based on all the time I'd spent researching, preparing, and giving the evidence.

Soon a company spy showed up in my office. I knew anything I said would be carried straight to the top, so I mentioned, "just between you and me," that I was in a position to reveal infractions committed by the local administration that were more serious than their charge against me. The next day I got a call from the director: There was no audit being planned; that was just a rumor; it was all a big mistake.

Soon I got word that my funding ploy had worked. The rehabilitation proposal was approved in routine fashion, and we were in business through 1979. I'm sure there was hell to pay at VACO when the phantom leader of the posse found that one of the director's underlings had let us slip the noose. However, I knew there was no way we'd ever finagle another grant, and I told everybody to start lining up another job well in advance. As for me, I'd decided to retire as soon as I became eligible in 1978. I was tired and discouraged. There was almost no interest in the evidence I'd accumulated for the DC perineural system, and I was pretty sure I'd never be able to do any research after this latest grant ran out.

Then a magazine article provided some support that induced me to continue for a while. In 1976, *Smithsonian* ran a piece called "If a Newt Can Grow a New Limb, Maybe We Can." The author, Robert Bahr, had written a popularized but accurate account after looking through the scientific literature and finding, among others, our papers on rat limb regeneration. Shortly afterward, I received a call from Don Yarborough, a congressional lobbyist for the American Paralysis Association, a group of people who were dissatisfied with the then-prevalent idea that nothing could be done for paraplegics. Obviously, spinal cord regeneration was the ultimate answer if it was possible. Yarborough asked if what Bahr had written was true.

When I said it was, he enthusiastically asked for more details. I told him the problems we'd encountered, both scientific and political, and the fact that I was soon to retire and disband the lab. He asked me to make no final decisions until he brought this situation to the attention of his contacts in Congress.

Shortly thereafter, I was asked to see Senator Alan Cranston, then chairman of the Senate Committee on Veterans' Affairs. Steve Smith was also invited. Also present were representatives from NIH and other agencies, who soon established their position. If the senator wanted to support crackpot stuff like this, they fumed, he would have to see to it that additional money was given to NIH, since they wouldn't fund this work at the expense of "good, solid research projects."

However, Cranston obviously was interested, and he must have leaned on the VA a little, for its policy changed briefly. The administrators expanded rehabilitation research, appointing a well-known orthopedic surgeon named Vernon Nickel to head it. As soon as he hit Washington, he called and asked what he could do for me. I told him that for years I'd been wanting to organize an international symposium on mechanisms of growth control and their clinical promise. Before I'd even finished describing the idea he asked how much I needed. I said $25,000, and he said it was on its way. Predictably, Vernon didn't last long in the capital; he was gone right after I organized the conference, but by then it was too late to cancel the funds.

The meeting was held in September 1979, and it exceeded all expectations. Every important researcher in the field was there, except for Meryl Rose, who can't stand crowds, and Marc Singer, who was ill. It brought together in one set of proceedings irrefutable evidence that a knowledge of the bioelectricity of growth would lead to incredible breakthroughs in medicine. Since then, a few more grants for regeneration work have opened up, and articles on bioelectromagnetism have begun to appear more often in the journals.

At a June 1978 meeting in Washington, requested by Senator Cranston to plan more regeneration research, my colleagues and I gave (we thought) an exciting and scientifically successful preview of the work that would be presented at the full-scale symposium a year later. At the end, however, the director stood up and said our work was largely erroneous and wholly without significance. "We see absolutely no reason to change the direction of VA research," she pontificated. "We see no reason to expand any programs in the area of regeneration." That was that for Cranston's initiative. The meeting appeared to be just a setup so the VA bigwigs could write him a letter saying they'd duly considered the work of all the experts, weighed it in the balance, and found it wanting.

However, my co-workers and I decided to give it the old college try for more funds. Our grant expired December 31, 1979, and we had to submit our new proposal by the preceding January. By now our nemesis was assistant chief medical director, so it was impossible to sneak one around her. The VA seems to change its application instructions every year, and this year we were specifically told our program was a "type III." Therefore we weren't to submit a detailed write-up of each proposed experiment. By this time we were obviously competent to do them, and we would be judged on the basis of past productivity. On that score we had no worries. We'd led the way in bioelectricity research and we were arranging the first conference ever in our specialty of growth control. Moreover, after our application was filed, during the symposium, I arranged for the chiefs of the various VA research sections to visit our lab, inspect the facilities, meet my replacement (Dave Murray), and discuss our future course. They said they wanted me to continue part time. Suddenly I felt we might survive after all.

Then, around Thanksgiving, came word that our application had been disapproved. The director had pulled the same switch. She'd changed procedures after receiving the proposal, instructing the reviewers to consider it as a "type I" application from a new investigator, with emphasis on *future* plans. As might be expected, one of the peers said, "This is a poorly written, overly ambitious, incompletely detailed proposal." Another wrote, "Dr. Becker is one of the pioneers in the field of electrically induced osteogenesis and regeneration. His work is visionary and exciting, yet at the same time it is controversial and lacks quantitation." The same old gobbledygook! Overly ambitious? By whose standards? Visionary and exciting? Exactly what do you want in a research project? Quantitation? If you can start to grow a leg back on a rat, what statistics do you want besides the procedures, photos, and the number of experimental animals and controls?

Soon after the rejection, I heard hints through intermediaries that the director might be willing to let us keep the laboratory together if I severed all my connections with it and if the remaining members agreed not to do any research on regeneration or electropollution. With such an agreement, it might be possible to get interim funds pending action on new proposals from my co-workers. The restrictions didn't leave much room to work, but at least my people's livelihood would be spared for a while, so we got back on the peer review treadmill.

Andy Marino, Joe Spadaro, and Dave Murray all submitted proposals of their own late in 1979. Approval of any would have kept the lab open, but the VA director went outside normal procedures to ensure rejection of each one. She chose an *ad hoc* committee instead of the usual peers to evaluate Spadaro's application. One of the critiques objected to a lack of guidance for him due to the fact that "we understand Becker is retiring." To judge Marino's plan for testing the three methods of electrical osteogenesis for long-term side effects, she bypassed the reviewers who would have been chosen routinely, picking instead someone who had no training in orthopedics but who could be counted on to reject anything remotely associated with me.

This man, a well-connected Purdue University embryologist, had for several years been making it a habit to ridicule my work in some of his own papers, while using it without credit as the basis of his own in other publications. In 1978, in response to a *Saturday Review* article describing regeneration work by me and others, a member of his lab had written a long, vituperative letter to the editor accusing me of "bad science" that had "made life difficult" for real researchers like him and his partners. He accused me of falsifying data on the rat limb experiment. He said the results I'd reported in three days were impossible, even though he'd never bothered to repeat the experiment himself. He ridiculed a claim that blastemas arose from *white* blood cells, an assertion that neither I nor anyone else had ever made. He accused me of misquoting other scientists' work to support my own ends. He charged me and Stephen Smith with trying to pass off a photo of an intact frog as one that had regrown a complete leg in one of Smith's experiments. Finally, he castigated me for avoiding scientific journals in favor of publishing my results in the popular press, even though I'd had well over a hundred papers printed in the peer-reviewed literature.

Steve Smith was moved to answer the letter point by point, closing with a denunciation of establishment scientists who responded to new ideas with ridicule and slander. Furthermore, he concluded, "I do not understand the rationale of those who feel that research is some obscure process whose results should be made known to the public only as a series of miraculous revelations. The decision as to what kind of research to support is essentially a political one in this day and age, and I firmly believe that an informed public opinion is a much better basis for that decision than general obfuscation."

Normally even the most scurrilous charges are thrashed out in the technical journals, however inadequate that process may be. The letter writer, however, took it upon himself to send a copy to one of the overseers of NIH funding. I suppose I could consider myself lucky that by then there was no more damage anyone could do me in that quarter.

The eminent embryologist's vendetta continued into 1980. In February of that year, the Purdue Office of Public Information issued a press release lauding the great man's own research and presenting him as a white knight doing single combat against "myths." In it he accused me of "plain ordinary fraud," repeating the accusations from his assistant's letter. As a result, I sent the president of Purdue a long, detailed letter itemizing his employee's actions during the past several years, and threatening to sue both him and the university. Immediately I got a call from the embryologist, in which he claimed he never *meant* anything like that, and asking what I planned to do. I asked to see some apologies in writing, and soon I received a nice letter expressing "deep regrets" for the "rather garbled" press release and belatedly acknowledging that he would never have done any regeneration work without the stimulus of Smith's experiments, which in turn had descended from mine.

The point of summarizing this embarrassing behavior is that, at about the same time he was talking with the Purdue press office, this gentleman was also writing the "impartial" review of Andy's proposal. We know who did the job because, in the small family of regeneration workers, the fellow's contentious style and his practice of quoting mainly his own research are immediately recognizable. His critique repeated some of the same old charges against my own work, even though the proposal wasn't mine. At one point he objected that one of Andy's experimental controls was to use the devices on healthy bone. He sniffed, "I am not aware that physicians are electrically stimulating uninjured bone in humans." Of course they weren't, but that part of the procedure was the perfect way to test for side effects in normal bone—the whole point of the experiment. The reviewer also wondered why some of the animals were to be sacrificed after one or two months while looking for long-term effects, knowing as well as any researcher that this was the normal way of following changes in tissue. After all this, he complained that there were *no* controls, when in reality the controls were to consist of animals allowed to live out their normal life-spans (which he conveniently overlooked), as well as the very procedures he objected to.

Several paragraphs of the pink sheets were deleted. Perhaps they contained more vitriol than even the VA thought was suitable for me. Another revealing criticism was left in, however. Since Andy had been part of my lab, the reviewer felt that I "would exert considerable influence on the character of the research proposed here," and for him that was an unacceptable contamination.

The denouement was completely predictable. All the proposals were rejected. We continued to work through 1980, supported at a low level of productivity by the interim funds. I kept a small part of the tissue culture lab going with a little money from the company that had made the black boxes for our bone-healing stimulators. In it we did a scaled-down version of Andy's proposed experiment to test for stimulation of malignant cells from electrical osteogenesis—and found it.

From January through June of that year, most of our energy went into drafting one last proposal and trying to get a fair hearing from someone else in the VA. This time we stipulated right in the proposal that I would officially retire and that Dave Murray would become the lab's principal investigator. Of course, the research director knew that I would still talk to these people, and on December 19 we were informed by phone of the final repudiation. As a last-ditch effort, I wrote a letter of appeal to the chief VA administrator, who'd shown favorable interest in our work a couple of years before. I asked him for a hearing and investigation, but to no avail.

The lab ceased to exist on New Year's Day 1981. The local VA chief had the gall to offer Andy and Joe jobs as night administrative officers. Instead, Marino went to work at Louisiana State University Medical School in Shreveport, where he still investigates the positive silver technique on a small scale. Spadaro also remained in orthopedic surgery, at the SUNY Upstate Medical Center right there in Syracuse. Maria Reichmanis decided she'd had enough of professional science, quit research entirely, and got married. This was the end of the foremost group working toward limb and spinal cord regeneration and one of the few bioelectromagnetism labs outside the DOD-industry orbit.

I've taken the trouble to recount my experience in detail for two reasons. Obviously, I want to tell people about it because it makes me furious. More important, I want the general public to know that science isn't run the way they read about it in the newspapers and magazines. I want lay people to understand that they cannot automatically accept scientists' pronouncements at face value, for too often they're self-serving and misleading. I want our citizens, nonscientists as well as investigators, to work to change the way research is administered. The way it's currently funded and evaluated, we're learning more and more about less and less, and science is becoming our enemy instead of our friend.

Glossary

Apatite: The mineral fraction of bone, microscopic calcium phosphate crystals deposited on the preexisting collagen structure of the bone, making it hard. See also COLLAGEN.

Axon: The prolongation of a nerve cell that carries a message, or stimulus, *away* from the cell body. For example, a motor nerve cell axon carries a contraction stimulus to a muscle. See also DENDRITE, NEURON.

Base pair: An association between two of the four fundamental chemical groups that make up all DNA and RNA molecules. Base pairs are the smallest structures that form units of meaning in the genetic code. The more base pairs, the larger the molecule.

Biological cycles: Changes in the activities of living things in an ebb-and-flow pattern. Such changes occur in almost all physical aspects, including sleep-wakefulness, hormone levels, and numbers of white cells in the blood. The predominant pattern is approximately twenty-four hours and usually follows the lunar day closely. See also CIRCADIAN RHYTHM.

Blastema: The mass of primitive, unspecialized cells that appears at the site of an injury in animals that regenerate. Blastema cells specialize and form the replacement part.

Circadian rhythm: The predominant biological cycle of all living things, from Latin *circa* ("approximately") and *dies* ("a day"). See also BIOLOGICAL CYCLES.

Collagen: A protein that makes up most of the fibrous connective tissue that holds the body's parts together. Tendons, ligaments, and scar tissue are composed almost entirely of collagen. It also forms the basic structure of bone. See also APATITE.

Crystal lattice: The precise, orderly arrangement of atoms in a crystal, forming a netlike structure.

Dedifferentiation: The process in which a mature, specialized cell returns to its original, embryonic, unspecialized state. During dedifferentiation the genes that code for all other cell types are made available for use by de-repressing them. See also DIFFERENTIATION, GENE, REDIFFERENTIATION.

Dendrite: The prolongation of a nerve cell that carries a message, or stimulus, *toward* the cell body. For example, sensory nerve cell bodies receive stimuli from receptors in the skin via their dendrites. See also AXON, NEURON.

Differentiation: The process in which a cell matures from a simple embryonic type to a mature, specialized type in the adult. Differentiation involves restricting, or repressing, all genes for other cell types. See also DE-DIFFERENTIATION, GENE, REDIFFERENTIATION.

DNA: The molecule in cells that contains genetic information.

Ectoderm: One of three primary tissues in the embryo, formed as differentiation (cell specialization) is just beginning. The ectoderm gives rise to the skin and nervous system. See also ENDODERM, MESODERM.

Electrode: A device, usually metal, that connects electronic equipment to a living organism for the purpose of measuring electrical currents or voltages in the organism, or delivering a measured electrical stimulus to the organism.

Electrolyte: Any chemical compound that, when dissolved in water, separates into charged atoms that permit the passage of electrical current through the solution.

Embryogenesis: The growth of a new individual from a fertilized egg to the moment of hatching or birth.

Endoderm: One of three primary tissues in the embryo, formed as differentiation (cell specialization) is beginning. It forms the digestive organs. See also ECTODERM, MESODERM.

Epidermis: The outer layer of skin, having no blood vessels.

Epigenesis: The development of a complex organism from a simple, undifferentiated unit, such as the egg cell. It is the opposite of preformation, in which a complex organism was thought to develop from a smaller, but similarly complex, antecedent, such as the homunculus that some early biologists thought resided in the sperm or egg cell.

Epithelium: A general term for skin and for the lining of the digestive tract.

Exudate: Liquid, sometimes containing cells, that diffuses out from a wound or surface structure of a living organism. Examples are a wound exudate and the slime exudate from the skin of fish.

Galvanotaxis: The movement of a living organism toward or away from a source of electric current.

Gene: A portion of a DNA molecule structured so as to produce a specific effect in the cell.

Gene expression: Specific structure and activity of a cell in response to a group of genes coded for such activity. For example, genes coded for muscle cause a primitive cell to assume the structure and function of a muscle cell.

Glia: A tissue composed of a variety of cells, mostly glial cells, that makes up most of the nervous system. These cells have been considered nonneural in the sense that they cannot produce nerve impulses. Therefore they have been thought unable to transmit information, rather having protective and nutritive roles for the nerve cells proper. This concept is changing. It is now known that glial cells have electrical properties that, while not the same as nerve impulse transmission, enable them to play a role in communication in the body.

Hertz: Cycles per second, the unit for measuring the vibratory rate of electromagnetic radiation. Named for Heinrich Hertz, German physicist who made the first experimental discovery of radio waves in 1888.

Homeostasis: The ability of living organisms to maintain a constant "internal environment." For example, the human body maintains a constant amount of dissolved oxygen in the blood at all times by means of various mechanisms that sense the oxygen level and increase or decrease the breathing rate.

In vitro: An experiment done in a glass dish on part of a living organism.

In vivo: An experiment done on an intact, whole organism.

Magnetosphere: The area around the earth in which the planet's magnetic field exerts a stronger influence than the solar or interplanetary magnetic field. It extends some 30,000 to 50,000 miles from the earth's surface. A prominent feature of the magnetosphere is the Van Allen belts, areas of charged particles trapped by the earth's magnetic field.

Magnetotactic: Active movement toward a magnetic pole.

Mesoderm: One of three primary tissue layers in the embryo, which develop as differentiation (cell specialization) begins. It becomes the muscular and circulatory system in the adult.

Mitosis: Cell division. Actual division takes only a few minutes but must be preceded by a much longer period during which preparatory events, such as duplication of DNA, take place. The entire process generally takes about one day.

Neoblast: An unspecialized embryonic cell retained in the adult bodies of certain primitive animals and called to the site of an injury to take part in regenerative healing.

Neuroepidermal junction: A structure formed from the union of skin and nerve fibers at the site of tissue loss in animals capable of regeneration. It produces the specific electrical currents that bring about the subsequent regeneration.

Neurohormone: A chemical produced by nerve cells that has effects on other nerve cells or other parts of the body.

Neuron, or neurone: A nerve cell.

Neurotransmitter: A chemical used to carry the nerve impulse across the synapse.

Osteoblast: A cell that forms bone by producing the specific type of collagen that forms bone's underlying structure.

Osteogenesis: The formation of new bone, whether in embryogenesis, postnatal development, or fracture healing.

PEMF: Pulsed electromagnetic field.

Periosteum: A layer of tough, fibrous collagen that surrounds each bone. It contains cells that turn into osteoblasts during fracture healing.

Photoelectric material: A substance that changes light into electrical energy, producing an electric current when light shines on it.

Piezoelectric material: A substance that changes mechanical stress into electrical energy, producing an electrical current when deformed by pressure or bending.

Potential: Another term for voltage, which may at times be limited to a voltage that exists without a current but is potentially able to cause a current to flow if a circuit is completed.

Preformation: See EPIGENESIS.

Pyroelectric material: A substance that changes thermal energy into electrical energy, producing an electric current when heated.

Redifferentiation: The process in which a previously mature cell that has dedifferentiated becomes a mature, specialized cell again. See also DEDIFFERENTIATION, DIFFERENTIATION.

Salamander: Any of a group of amphibians related to frogs but retaining a tail throughout their lives. Salamanders live in water or moist environments. Most are 2 to 3 inches long, but some grow to more than a foot in length. Since salamanders are vertebrates, with an anatomy similar to ours, and since they regenerate many parts of their bodies very well, they are the animals most commonly used in regeneration research.

Schwann cells: The cells that surround all of the nerves outside of the brain and spinal cord. See also GLIA.

Sciatic nerve: The main nerve in the leg. It includes both motor nerve fibers carrying impulses to the leg muscles and sensory nerve fibers carrying impulses to the brain.

Semiconduction: The conduction of electrical current by the movement of electrons or "holes" (the absence of electrons) through a crystal lattice. It is the third and most recently discovered method of electrical conduction. The others are metallic conduction, which works by means of electrons traveling along a wire, and ionic conduction, which works by movement of charged atoms (ions) in an electrolyte. Semiconductors conduct less current than metals but are far more versatile than either of the other types of conduction. Thus they are the basic materials of the transistors and integrated circuits used in most electronic devices today.

Superconduction: The conduction of an electrical current by a specific material that under certain circumstances (generally very low temperatures) offers no resistance to the flow. Such a current will continue undiminished as long as the necessary circumstances are maintained.

Synapse: The junction between one nerve cell and another, or between a nerve cell and some other cell. See also NEUROTRANSMITTER.

Undifferentiated: Unspecialized, a term applied to cells that are in a primitive or embryonic state. See also DEDIFFERENTIATION, DIFFERENTIATION, REDIFFERENTIATION.

Vertebrate: Any of the animals that have backbones, including all fish, amphibians, reptiles, birds, and mammals. All vertebrates share the same basic anatomical arrangement, with a backbone, four extremities, and similar construction of the muscular, nervous, and circulatory systems.

INDEX

ABOUT THE AUTHORS

Robert O. Becker, M.D., is a pioneer in the field of research on regeneration and its relationship to electrical currents in living things. He began his career as an orthopedic surgeon in 1956 at the VA hospital in Syracuse, New York, while teaching at the Upstate Medical Center. He lives in upstate New York.

Gary Selden is a writer who specializes in medical and scientific topics. He has written articles for *Science Digest, Ms., Cosmopolitan,* and many other magazines.